Nanocomposite Materials: Synthesis, Properties and Applications

Nanocomposite Materials: Synthesis, Properties and Applications

Tristan Carson

MURPHY & MOORE
www.murphy-moorepublishing.com

Nanocomposite Materials: Synthesis, Properties and Applications
Tristan Carson
ISBN: 978-1-63987-380-7 (Hardback)

MURPHY & MOORE

Published by Murphy & Moore Publishing,
1 Rockefeller Plaza,
New York City, NY 10020, USA

Cataloging-in-Publication Data

Nanocomposite materials : synthesis, properties and applications / Tristan Carson.
p. cm.
Includes bibliographical references and index.
ISBN 978-1-63987-380-7
1. Nanocomposites (Materials). 2. Composite materials. 3. Nanostructured materials. I. Carson, Tristan.
TA418.9.N35 N36 2022
620.5--dc23

For more information regarding Murphy & Moore Publishing and its products, please visit the publisher's website www.murphy-moorepublishing.com

TABLE OF CONTENTS

PREFACE

This book aims to help a broader range of students by exploring a wide variety of significant topics related to this discipline. It will help students in achieving a higher level of understanding of the subject and excel in their respective fields. This book would not have been possible without the unwavered support of my senior professors who took out the time to provide me feedback and help me with the process. I would also like to thank my family for their patience and support.

The multiphase solid material which has one, two or three dimensions of less than 100 nanometers are known as nanocomposites. It also includes structures that have nano-scale repeat distances between the various phases that form the material. The underlying goal of this field is to make materials that are more flexible and have improved mechanical, electrical, optical, catalytic, thermal, electrochemical and physical properties as compared to the component materials. It includes the study of colloids, porous media, gels and copolymers. These materials have high surface to volume ratio of the reinforcing phase. They can be divided into four types. Some common types of composite materials are metal-matrix nanocomposites, magnetic nanocomposites, ceramic-matrix nanocomposites and polymer-matrix nanocomposites. The topics covered in this extensive book deal with the core aspects of nanocomposite materials. It studies, analyzes and upholds the pillars of nanocomposite materials and its utmost significance in modern times. For someone with an interest and eye for detail, this book covers the most significant topics related to this field.

A brief overview of the book contents is provided below:

Chapter - Introduction

Multiphase solid materials that consist of phases in which, one of the phases have one, two or three dimensions at nano-scale repeated distances are termed as nanocomposites. It differs from conventional composite materials. This chapter has been carefully written to provide an easy understanding of nanocomposite materials.

Chapter - Polymer Nanocomposites

The polymers that have nanoparticles and nanofillers dispersed in the polymer matrix are called polymer nanocomposites. It improves the performance properties of the polymer and has various biomedical and electrical applications. This chapter closely examines polymer nanocomposites to provide an extensive understanding of the subject.

Chapter - Ceramic Nanocomposites

The nanocomposites that comprise of more than one solid phase in which at least one phase has dimensions in the range of 50nm to 100nm are called ceramic nanocomposites.

A few of its examples are polymer-ceramic nanocomposite, alumina ceramic nanocomposite, chromium carbide ceramic nanocomposite, etc. This chapter discusses the subject of ceramic nanocomposites in detail.

Chapter - Nanoporous Materials

The organic or inorganic materials comprising of pores of size less than 100nm that support the porous structures are termed as nanoporous materials. They can be branched into three categories - microporous materials, mesoporous materials and macroporous materials. The topics elaborated in this chapter will help in gaining a better perspective of nanoporous materials.

Chapter - Metal Matrix Nanocomposites

The nanocomposites which consist of metal as the matrix, and ceramic or organic compounds as the reinforcement are called metal matrix nanocomposites. These can be classified into continuous and non-continuous reinforced materials. This chapter delves into the subject of metal matrix nanocomposites for an in-depth understanding of the subject.

Tristan Carson

Introduction 1

- Nanocomposite
- Applications of Nanocomposite Material

Multiphase solid materials that consist of phases in which, one of the phases have one, two or three dimensions at nano-scale repeated distances are termed as nanocomposites. It differs from conventional composite materials. This chapter has been carefully written to provide an easy understanding of nanocomposite materials.

Nanocomposite

A nanocomposite is a composite material, in which one of the components has at least one dimension that is nanoscopic in size that is around 10^{-9} m. Scaling might be helpful: A coin is on order of 1-2 mm thick, or 10^{-3} m; a carbon fibre, commonly used as a reinforcement in sporting goods, is approximately 7 μm in diameter, or 10^{-6} m; a carbon-carbon chemical bound, the basic unit of life, is about 1.5 Å, or 10^{-10} m.

A composite/Nanocomposite material is constituted by a matrix and a reinforcement consisting of fibers. The matrix itself comprises a resin and filler, the goal of which is to improve the characteristics of the resin while reducing the production cost. From a mechanical point of view the filler-resin system behaves as a homogeneous material, and the composite is considered as being made of a matrix and reinforcement.

The reinforcement brings to the composite material its greater mechanical performance, whereas the role of the matrix is to transmit to the fiber to the external mechanical load and protect the fibers against external attack.

The type of the reinforcement matrix association depends upon the constraints imposed on the designer: High mechanical characteristics, good thermal stability, cost, resistance to corrosion, etc.

Most nanocomposites that have been developed and that have demonstrated technological importance have been composed of two phases, and can be microstructurally classified into three principal types:

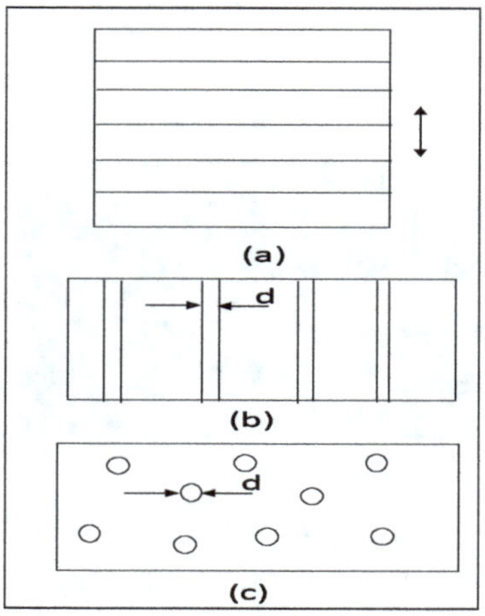

(a) Nanolayered composite composed of alternating layers of nanoscale dimension; (b) nanofilamentary composites composed of a matrix with embedded (and generally aligned) nanoscale diameter filaments; (c) nanoparticulates composites composed of a matrix with embedded nanoscale particles. As with conventional composites, the properties of nanocomposites can display synergistic improvements over those of the component phases individually.

However by reducing the physical dimensions(s) of the phase(s) down to the nanometer length scale, unusual and often enhanced properties can be realized. An important microstructural feature of nanocomposites is their large ratio of interphase surface area to volume.

Classification of Nanocomposites

On the bases of their engineering applications, nanocomposites can be classified either, Functional materials i.e. based on electrical, megnetical, and/or optical behaviour, example is nanolayered semiconductor (semiconductor superlattice) composed of alternating layer of single crystal GaAs and Ga Al_xAs_{1-x}.

Structural materials i.e. based on their mechanical properties.

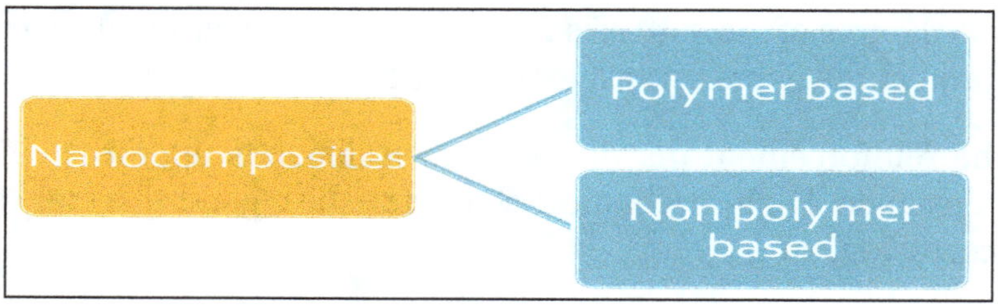

1. Non-polymer based Nanocomposites:

Non-polymer based nano composites can further be classified as below:

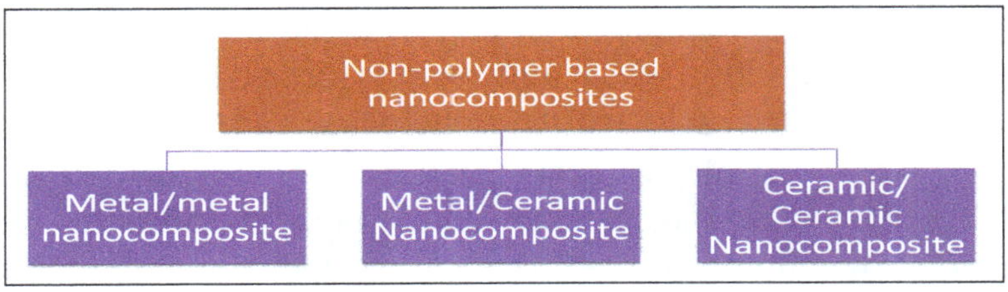

2. Polymer based Nanocomposite:

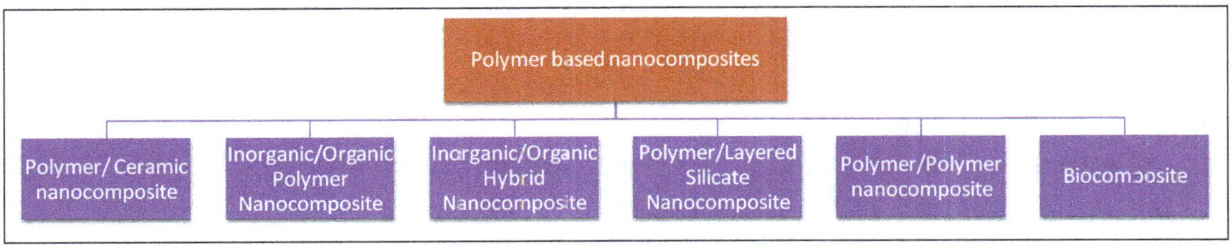

Depending on the type of filler, i.e., the nanoscale material of the nanocomposites, for sensing applications they are divided into:

- Metal oxide–metal oxide–based nanocomposites.
- Polymer-based nanocomposites.
- Carbon-based nanocomposites.
- Noble-metal–based nanocomposites.

Due to their large aspect ratios (i.e., size-to-volume ratios), submicrometer size, and unique properties, nanosensors, nanoprobes, and other nanosystems are revolutionizing the fields of chemical and biological analysis. Catalysis, separation, sorption, and fuel cells are other important fields for nanocomposite applications. Nanocomposites can be considered as solid structures with nanometer- scale dimensional repeat distance between the different phases. Typically, nanocomposites are classified as inorganic matrix (inorganic-inorganic nanocomposites), organic filler in organic (organic-organic nanocomposites), and hybrid materials, i.e., organic in inorganic or inorganic in organic matrix.

General Approaches to Nanocomposite Fabrication

Nanostructures of different materials for fabrication of nanocomposites can be prepared by various techniques. For example, nanocomposite systems such as metal nanoparticles on ceramic supports can be prepared simply by evaporating metal onto the chosen substrate or dispersal via solvent chemistry. On the other hand, nanocomposites that have complex structures with coexisting ceramic and polymeric phases are difficult to prepare using these methods, and therefore the preparation of such materials requires novel processing techniques such as template synthesis, scanning probe electrochemical methods, electrospinning, etc. In the development of nanostructured materials, a template is defined as a central structure within which a network is formed in such a way that removal of the template creates a filled cavity with morphological and/or stereochemical features related to the template. Template synthesis entails the preparation of a variety of micro- and nanomaterials of a desired morphology and therefore provides a route to enhancing nanostructure order. In nontemplate selfassembly, the individual components interact to produce a larger structure without the assistance of external forces or spatial constraints.

Despite the variety of approaches to synthesizing different nanostructures, there is a need for a method capable of making pure, uniform, template-free nanostructures. Such a fabrication strategy requires only the mixing of components to achieve an ordered structure and is appealing both for its simplicity and its potential efficiency. Template-free methods of synthesizing nanostructures have several advantages, including simple synthesis and purification with no template-removing steps needed. Also, uniform nanofibers are formed, which are easily scalable and reproducible. They show superior performance as sensors because the diameter of nanomaterials is at nanoscale and they are water-dispersible, which facilitates environmentally friendly processing and biological applications.

Non-Polymer based Nanocomposites

Non- polymer based nanocomposite materials can be classified as follows:

Metal/Metal Nanocomposite

Bimetallic nanoparticles either in the form of alloy or core-shell structures or being investigated in some depth because of their improved catalytic properties and changes in the electronic/optical properties related to individual, separate metals. It is postulated their interesting Physico-chemical properties, result from the combination of two kinds of metals and their fine structures.

Metal/Ceramic Nanocomposites

In these types of composites, the electric, magnetic, chemical, optical and mechanical properties of both phases are combined. Size reduction of the components to the nanoscale causes improvement of the above mentioned properties and leads to new application. The polymer precursors techniques offers an attractive rough to such composites proving a chemically inert and hard ceramic matrix.

Ceramic/Ceramic Nanocomposites

Ceramic Nano composites could solve the problem of fracture failures in artificial joint implants; these would extend patient's mobility and eliminate the high cost of surgery.

The use of Zirconia-toughened alumina nanocomposite to form Ceramic/ceramic implants with potential life spends of more than 30 years.

Polymer based Nanocomposites

Polymer nanocomposites are composites with a polymer matrix and filler with at least one dimension less than 100 nm. The fillers can be (clay), high aspect ratio, nanotubes and Lower aspect ratio or Nanoparticles.

Polymer/Ceramic Nanocomposite

Nanocomposites consist of single ceramic layers(1nm thick) homogeneously dispersed in a continuous matrix. The host ceramic layer tend to orient themselves parallel to each other due to dipole-dipole interaction.

Natural Bone is a nanocomposite-bone consisting of approximately 30% matrix (collagen) material and 70% nanosized minerals (hydroxyapatite).

Inorganic/Organic Polymer Nanocomposites

Metal polymer nanocomposites attract attention because of unique properties of metal clusters which are dispersed in polymer matrix. the typical size of such metal cluster is approximately 1- 10 nm. The properties of clusters and nanoparticles (band gap, spectral properties, the transport of electrons)are very different from those of bulk materials and from individual atom or molecules.

The size and grains depends on mobility of the metal atoms on the polymer surface. For example, in the case of polymethylmethacrylite (PMMA) polymer the cluster size depends on the amount of the cross linking of the polymer, which obviously changes the mobility of the metal atoms.

Inorganic/Organic Hybrid Nanocomposite

Hybrid inorganic/organic materials are not simply physical mixtures; they can be broadly defined as nanocomposites with organic and inorganic components intimately mixed. Indeed, hybrids are either homogenous system derived from monomers and miscible organic/inorganic components, or heterogenous systems (nanocomposites) where at least one of the component has the scale of nanometer.

Polymer/Layered Silicate Nanocomposites

Polymer/Layered silicate (PLS) nanocomposites materials are attracting considerable interest in polymer science research. In recent years the PLS nanocomposites have attracted great interest both in industry and academia, because they often exhibits remarkable improvements in materials when compare with virgin polymer and conventional macro and macro composites. Hactorite and montmorillonite are among the most commonly used smectite-type- layered silicates for the preparation of the nanocomposites.

Polymer/Polymer Nanocomposites

Polymers are more than ever under pressure to be chip and offered property profiles. The gap between block co-polymer self assembly and offer nanostructured plastic endowed with still unexplored combinations of properties is getting narrower. Mixtures of different polymers often phase separate,even when their monomer mixed homogenously.

Biocomposites

Metals and metal alloys are used in orthopaedics, dentistry and other load bearing applications. Ceramics are used with emphasis on either their chemically inert nature or high bioactivity; all polymers are used for soft tissue replacements and used for many other non structural application.

Naturally occurring composites are within us all. Collagen is highly abundant and varies with more than 14 types discovered. All variations are formed from tropo-collagen molecules, which are inelastic.

Classification of Nanocomposite on the bases of Rainforcememnt Material

Metal Oxide–based Nanocomposites

Nanocomposites can be based on a metal oxide matrix in which the filler is also metal oxide nanoparticles, nanowires, etc. As is by now well known, metal oxides are important semiconductors which can be used as sensing materials in chemical sensors. The advantages offered by wide-band-gap semiconductor oxides include their stability in air, relative inexpensiveness, and easy preparation in the ultradispersed state. Since porous metal oxides have attractive properties such as simplicity of preparation, tunable porosity, good chemical stability, low-temperature encapsulation, negligible swelling, mechanical and biodegradable stability, and high sensitivity at lower operating temperatures for detection of reducing and oxidizing gases, they have been used for the fabrication of chemical sensors and biosensors.

Metal oxide- based nanocomposites can be prepared by various methods. There are some methods like mechanical and chemical, which can be used. By using mechanical method which can grind the metal and give very small grain size and also give homogeneous mixture (ball milling). In this process, alloying occurs as a result of repeated breaking up and joining (welding) of the component particles. The process can prepare highly metastable structures such as amorphous alloys and nanocomposite structures with high flexibility. Materials prepared by this method were used in a gassensing application. For example, it was found that sensors based on Fe_2O_3 (Sn, Ti, Zr) nanocomposites prepared this way showed improved sensitivity to ethanol and hydrocarbons. Scaling up of synthesized materials to industrial quantities is easily achieved for this mechanical alloying process, but purity and homogeneity of structures produced remains a challenge. In addition to erosion and agglomeration, highenergy milling can provoke chemical reactions, which can influence the properties of nanocomposites. Metal oxides–based nanocomposites can also be prepared by sol-gel processes. Aerogels, because of their high-porosity structure, are an ideal starting material for use in nanocomposites Aerogel nanocomposites can be fabricated in various ways, depending on when the second phase is introduced into the aerogel material. The second component can be added during the sol-gel processing of the metal oxides (before supercritical drying). It can also be added into the vapour phase (after supercritical drying), or chemical modification of the aerogel particles may be effected through reactive gas treatments. These general approaches can produce many varieties of nanocomposites. Deposition from the aerosol phase and laser ablation or pulsed laser deposition can also be successfully used for deposition of nanocomposites.

Polymer-based Nanocomposites

Polymer–nanoparticle composite materials have also attracted the interest of a number of researchers, due to their synergistic and hybrid properties. Ease of processability of organic polymers combined with the better mechanical and optical properties of nanoparticles has led to the fabrication of many devices. These are the nanocomposites based on polymer filler in any matrix, better described as nano filled polymer composites can be prepared using polymers. Polymer/ceramic nanocomposites (polymer matrices filled with ceramic nanopowders) are a promising material for embedded capacitors. They combine the high dielectric constant of ceramic powders and the processability and flexibility of polymers. In addition, advances in nanotechnology may enable polymer/metal nanocomposites (polymer matrices dispersed with metal nanopowders) to compete favourably with more traditional ceramic-filled polymer composites. (Nanocomposites: metal and ceramic filled polymer for Dielectrics, nanoEngineered Materials).

The most important step in fabrication of polymer-based nanocomposites is the dispersion of filler in the matrix. Various mechanochemical approaches, including sonication by ultrasound, can be used for this purpose. However, the scope of such approaches for dispersing the nanoparticles is limited by reaggregation of the individual nanoparticles and establishment of an equilibrium state under certain conditions, which determines the size distribution of the agglomerate of the dispersed nanoparticles. Other limitations are related to temperature conditions and the limited stability of some types of inorganic nanoparticles to mechanical impacts.

Particles coated with a polymer shell are considerably more stable against aggregation because of a large decrease of their surface energy in comparison with bare particles.

Such a polymer shell can be obtained by first synthesizing the inorganic nanoparticles and then dispersing them in a polymer solution. Finally, the polymer-coated inorganic nanoparticles are precipitated into a nonsolvating phase. This is the so-called ex-situ approach. Such a process of polymer shell formation on preformed inorganic cores can also be realized by polymerization of the desired monomer with organic nanoparticles dispersed in it. Then the nanocomposite material is formed. The ex-situ approach is the most general one because there are no limitations on the kinds of nanoparticles and polymers that can be used. The presence of such a shell increases the compatibility of the particles in the polymer matrix and makes it easier to disperse them.

In some cases, the process of polymer-based nanocomposite formation and nanoparticle preparation can be combined into one process or performed as a series of consecutive processes in one reactor (the in situ approach). In the in situ methods, nanocomposites are generated inside a polymer matrix by precursors, which are transformed into the desired nanoparticles by appropriate reactions. In situ approaches are currently getting much attention because of their obvious technological advantages over ex situ methods. One-step synthesis leads to improved compatibility of the filler and the polymer matrix and enhanced dispersion of the filler.

Up to now, in situ formation of particles in a liquid medium has probably been the most widely used method for the preparation of polymer nanocomposites containing isotropic inorganic particles. Commonly, soluble inorganic or organometallic compounds are converted by chemical reactions to colloids in water or organic solvents. The polymer may be already present during colloid synthesis or may be added afterwards. The particle dispersion can be destabilized or stabilized by the polymer, depending on the system. In the former case, the nanocomposite forms spontaneously by co-precipitation after colloid formation; in the latter case, nanocomposites can be obtained by addition of a solvent that acts as a co-precipitation agent, by casting followed by solvent evaporation, or by spin coating.

Inorganic particles can also be prepared in situ in solid polymer matrices, e.g., by thermal decomposition of incorporated precursors, reaction of incorporated compounds with gaseous species, or when polymer films containing an incorporated precursor are immersed in liquids containing the reactive species required for formation of the desired colloid. If solid reaction by-products arise from the particle synthesis, they can be embedded in the nanocomposites, hence the formation of volatile reaction side products which are able to leave polymer matrices should be preferred if possible.

Several powders of surface-modified colloids have been found to disperse well in liquids. Such dispersions can be mixed with dissolved polymers, and subsequently nanocomposites can readily be obtained by casting followed by solvent evaporation or by spin coating. The preparation of nanocomposites by diffusion of dispersed colloids in polymer films is also possible. This method is suited only for rather insoluble polymers with good swelling behaviour or for very thin films. Some nanoparticles which can be isolated as powders disperse in polymer melts without pronounced agglomeration of the primary particles, especially colloids that are coated with a layer of organic molecules. This affords a simple technique for the preparation of nanocomposites by direct mixing of particles and polymers.

Carbon Nanotube–based Nanocomposites

Among all carbon products, carbon nanotubes have been of great interest, both from a fundamental point of view and for potential applications. Their mechanical and unique electronic properties open up a broad range of applications, including nanoelectronic devices, composites, chemical sensors, biosensors, and more. They also provide tremendous opportunities in the design of multifunctional materials systems. In particular, they promise to provide solutions to many vexing problems encountered during the application of traditional composite materials. For example, they are electrically conductive and therefore are suitable for applications that require the ability to discharge electrostatic potentials.

Carbon nanotubes (CNTs) can be classified as single-walled nanotubes (SWNTs) and multiwalled Nanotubes (MWNTs). SWNTs consist of a cylindrical single sheet with a diameter between 1 and 3 nm and a length of several micrometers. They possess a cylindrical nanostructure formed by rolling up a single graphite sheet into a tube. MWNTs consist of a coaxial arrangement of concentric single nanotubes like rings of a tree trunk separated from one another by 0.34 nm. They usually have a diameter of about 2–20 nm. T e production of SWNTs or MWNTs is highly dependent on the synthesis process and conditions.

Synthesis

For processing nanotube composites, many approaches involve several steps that may include highenergy sonication, chemical polymerization of the corresponding monomer in the presence of CNTs, electrochemical synthesis of polymers on CNTs electrode, solution–evaporation processing, surfactantassisted processing through formation of a colloidal intermediate, functionalization of nanotubes with the polymer matrix, and high shear mixing. A commonly used solution–evaporation method for preparing nanotube–polymer composites involves mixing nanotube dispersions with a solution of the polymer and then evaporating the solvent in a controlled way. T e low viscosity of the polymer solution allows the nanotubes to move freely through the matrix. T e solution mixing approach is limited to polymers that dissolve readily in common solvents. An alternative is to use thermoplastic polymers (i.e., polymers that soften and melt when heated), and then apply melt processing techniques. Thus, shear mixing can be used to produce a homogeneous dispersion of nanotubes and extrusion to produce nanotube alignment or to fabricate objects in the required form by injection molding. Another alternative method for preparing nanotube–polymer composites is to use the monomer rather than the polymer as a starting material, and then carry out in situ polymerization.

The problem of insolubility of carbon nanotubes due to strong vander Waals attractions and chemical inertness hinders their uniform dispersion and incorporation in any matrix. A homogeneously dispersed filler in the polymer matrix reduces the possibility of nanotube entanglement, which can lead to significant changes in composite behaviour. The nanotube aggregation within a polymer system would certainly have a negative impact on its stiffening ability. As yet, the nature of these entanglements and their influence on the composite properties is a little-understood area. To overcome nanotube aggregation, functionalization of nanotubes is done with groups that facilitate their incorporation into a material through covalent bonding

Noble Metal based Nanocomposites

Many techniques for incorporating metal nanoparticles into the polymeric matrix have been published in the literature. Such combinations require blending or mixing the components with the polymer in solution or in melt form. However, conducting polymers are not fusable and are generally insoluble in common solvents. Therefore, synthesis techniques had to be developed to incorporate inorganic components into the conducting polymer. There are two main kinds of nanosized composites of conducting polymers with metals: Metal-core nanoparticles covered with a conducting polymer shell; and metal nanoparticles embedded into a conducting polymer matrix. Metal core nanoparticles covered with a conducting polymer shell are usually prepared by the chemical or electrochemical polymerization of a thin, nanometer-sized layer of a conducting polymer onto colloid metal particles. There are many techniques for the deposition of nanometer-sized conducting polymer layers onto different substrates, including nanosized ones. Metal nanoparticles embedded into a conducting polymer matrix can be easily obtained by the chemical reduction of metal ions from their salt solution at the conducting polymer/solution interface. Many conducting polymers, when present in their reduced form, have sufficiently high reducing power with respect to some metal ions; thus, some metal ions that have a relatively high positive redox potential, e.g., gold, silver, platinum, and copper, can be reduced at a layer of conducting polymer, composites with metal-core nanoparticles covered with a polymer shell can also be prepared by chemical or electrochemical polymerization of a thin, nanometer-sized layer of polymer onto colloidal metal particles. For example, the Au/PPy nanocomposite (inorganic filler in organic matrix) deposited onto a glassy carbon electrode via electrosynthesis shows that Au nanoparticles are uniformly distributed in the PPy composite film and the film exhibits a highly microporous structure with polymer fibrils spanning diameters in a micrometer range. It has been proposed that the mechanism of formation may be that the Au-nanoparticles act as nuclei for the pyrrole polymerization when they diffuse on to the electrode surface during electrosynthesis.

A novel concept of fabrication of multilayer network films on electrodes to form stable anionic monolayers (templates) on carbon and metals has been developed by Karnicka et al. In these hybrid films, the layers of negatively charged polyoxometallate or polyoxometallate-protected (stabilized) Pt nanoparticles are linked or electrostatically attracted by ultrathin layers of positively charged conducting polymers (PANI, PPy, PEDOT). The films are functionalized and show electrocatalytic properties toward reduction of nitrite, bromate, hydrogen peroxide, and oxygen. The size of noble metal particles in polymer-based nanocomposites can be varied from 5 to 400 nm depending on the method of nanocomposite preparation.

There are many routes for preparing metal/metal oxide–based nanocomposites as well. Ion exchange and wet impregnation are the simplest routes for introducing Pt, Pd, Ru, and Ag nanoparticles into a metal oxide porous matrix. However, ion exchange and wet impregnation often result in low guest dispersion because the metal salts can easily diffuse onto the outer surface of the host metal oxide matrix during the reduction or thermal treatment process, so large metal particles can readily form on the particle surface.

Applications of Nanocomposite Material

Nanocomposite materials have potential applications in wide range of fields which includes information and communication technologies, medical care, water decontamination, polymer electrolyte, agriculture and packaging etc.

Experimental work has generally shown that virtually all types and classes of nanocomposite materials lead to new and improved properties, when compared to their macrocomposite counterparts. Therefore, nanocomposites promise new applications in many fields such as mechanically-reinforced lightweight components, non-linear optics, battery cathodes and ionics, nanowires, sensors and other systems.

Such mechanical property improvements have resulted in major interest in nanocomposite materials in numerous automotive and general/industrial applications. These include potential for utilisation as mirror housings on various vehicle types, door handles, engine covers and intake manifolds and timing belt covers. More general applications currently being considered include usage as impellers and blades for vacuum cleaners, power tool housings, mower hoods and covers for portable electronic equipment such as mobile phones, pagers etc.

Nanomer nanoclays provide plastics product development teams with exciting new polymer enhancement and modification options. With the proper choice of compatibilizing chemistries, the nanometer-sized clay platelets interact with polymers in unique ways. Application possibilities for packaging include food and non-food films and rigid containers. In the engineering plastics arena, a host of automotive and industrial components can be considered, making use of lightweight, impact, scratch-resistant and higher heat distortion performance characteristics.

Oil and Gas Pipelines

Corrosion has a costly and deleterious effect on aging infrastructure throughout the world. As such, considerable attention has been focused on innovative techniques to arrest corrosion in the carbon steel found in pipelines, bridges, and water systems. In the United States, the annual cost associated with corrosion damage of structural components is greater than the combined annual cost of natural disasters, including hurricanes, storms, floods, fires and earthquakes. Similar findings have also been made by studies conducted in the United Kingdom, Germany, and Japan.

According to the U.S. Department of Transportation Office of Pipeline Safety, "between 1989 and 2008 pipeline corrosion incidents resulted in over $582M in property damages, 28 fatalities and 94 injuries." The need to manage and mitigate corrosion damage has therefore rapidly increased, as materials are placed in more extreme environments and pushed beyond their original design life.

Corrosion damage and failure are not always considered in the design and construction of pipeline systems. Even if corrosion is considered, unanticipated changes in the environment in which the structure operates can result in unexpected corrosion damage. Moreover, combined effects of corrosion and mechanical damage, and environmentally assisted material damage can result in unexpected failures due to the reduced load carrying capacity of pipelines.

The application of nanocomposites is becoming increasingly important in the manufacture and structural repair of damaged pipelines. In their work, Kessler and Goertzen pointed out that nanocomposites offer more advantages in pipelines manufacture, while its overwraps are used to repair corroded steel pipelines, as the repair can be completed in a relatively short period and the fluid transmission in the piping system can remain undisrupted while the repair is being made.

Fiber-reinforced nanocomposite pipelines are emerging as a feasible alternative to steel pipelines with regard to performance and cost. The pipeline is typically constructed including an inner nonpermeable barrier tube that transports the fluid (pressurized gas or liquid), a protective layer over the barrier tube, an interface layer over the protective layer, multiple glass or carbon fiber composite layers, an outer pressure barrier layer, and an outer protective layer. It is a nanocomposite structure in the purest engineering sense of the term, as each of the several components provides a distinct function and the interaction between the components produces a structure with exceptional performance characteristics.

The pipeline has improved burst and collapse pressure ratings, increased tensile strength, compression strength, and load carrying capacity, compared to non-reinforced, non-metallic pipelines. The ability of re-inforced nanocomposite piping to withstand large strains allows the piping to be coiled such that long lengths can be spooled onto a reel in an open bore configuration.

In addition, the pipe can be manufactured with fiber optics, copper signal wires, power cables or capillary tubes installed directly into the structural wall of the piping. This offers the option of manufacturing the pipe with embedded sensors and operating it as a socalled smart structure. Sensors embedded in the pipe can be powered via copper wire from remote locations and real-time data from the sensors can be returned through fiber optics. This provides the unique advantage of lifetime performance monitoring of the pipe.

According to Mays, the application of reinforced nanocomposites in oil and gas pipelines has the following advantages:

- Anisotropic characteristics of nanocomposite piping provide extraordinary burst and collapse pressure ratings, increased tensile and compressive strengths, and increased load carrying capacities.
- No welding and minimal joining - many miles of continuous pipeline can be emplaced as a seamless monolith.
- Emplacement requirements should be dramatically less than those for metal pipe, enabling the pipe to be installed in areas where right-of-way restrictions are severe.
- Structurally integrated sensors provide real-time structural health monitoring and could reduce need for pigging.
- Corrosion resistant and damage tolerant.
- Meets or exceeds published and consensus standards for pipeline in oil and gas applications.

The basic idea in implementing the nanocomposites in mechanical stream is the resistant to fracture and the often occurrence of wear and tear of the machine parts. Nanocomposites used as a blend against plastics can be used for strengthening the portions of the automobiles where higher efficiency is required.

As the world is affected by the pollution the automobile manufacturers are working towards developing a technology which controls the same cost effectively, this led to the acceptance of polymeric nanocomposites. Owing to their polymeric nature, polymer nanocomposites fit this description. Because of their nanometer size features, polymeric nanocomposites possess unique properties, such as enhanced mechanical, impact, barrier and heat resistant properties, compared to other composites.

Combining the unique properties of nanocomposite and recyclable polymers to produce light-weight recyclable and biodegradable polymer/ nanocomposite is a great challenge. These compositions were widely used for making the body parts of the automobiles. The industry was mainly concerned over the following aspects:

- Weight reduction.

- Improved performance.
- Aesthetics.
- And recyclability.

Nanotechnology is already driving changes throughout this industry at nearly every level involving material, components, and systems. This is because most cars produced in the United States contain some nanocomposite material, most typically carbon nanotube in nylon blend for the use of the fuel system to protect against static electricity.

According to Buchholz , a plastic nanocomposite is being used for "step assists" in the General Motors Safari and Astro vans as it is scratchresistant, light-weight, and rust-proof, and generates improvements in strength and reductions in weight, which lead to fuel savings and increased longevity. And in 2001, Toyota started using nanocomposites in a bumper that makes it 60% lighter and twice as resistant to denting and scratching.

Nanocrystals of various metals have been shown to be 100 percent, 200 percent and even as much as 300 percent harder than the same materials in bulk form. Because wear resistance often is dictated by the hardness of a metal, parts made from nanocrystals will significantly last longer than conventional parts.

Aircrafts

Researchers have made relatively awesome discoveries on nanocomposites over the last decade, ever since the pioneering work on nanocaly by the company Toyota. The dispersion of the silicate nanolayer with its high aspect ratio, large surface area, and high stiffness within a polymer matrix results in significant improvement of the properties of polymeric materials, including mechanical properties, barrier properties, resistance to solvent swelling, ablation performance, thermal stability, fire retardancy, controlled release of drugs, anisotropic electrical conductivity, and photo activity.

Layered-silicate nanocomposites have great applications, ranging from automotive and aerospace to food packaging and tissue engineering. Epoxy materials are widely used in adhesives, coatings,composites and electronics. These are also used in designing of aircraft parts too.

This epoxy system has a high glass transition temperature (Tg), good mechanical and physical performance characteristics, and low viscosity. In addition, epoxy nanocomposites as primer layer for aircraft coatings for improved anticorrosion properties are used.

High performance nanocomposites are also used in fuselage sinks in aircrafts.

Electronics

Conductive nanocomposites are capable of conducting electric current well owing to the electric charges in their structure. Polycarbonates which is an insulator can be made conductive Polycarbonates, the inexpensive plastics known for their excellent optical and mechanical properties, could in future, find applications into newer and more important horizons. Polycarbonates are tagged as poor electrical conductors, but a research team from University of Houston (UH) has altered this very property by adding carbon nanotubes to them thereby resulting in highly conductive nanocomposites. The team has come up with a strategy to achieve higher conductivities using carbon nanotubes in plastic hosts than what has been currently achieved. By combining nanotubes with polycarbonates, the team was able to reach a milestone of creating nanocomposites with ultra-high conductive properties. Shay Curran, associate professor of physics at UH demonstrated ultra-high electrical conductive properties in these plastics by mixing them with just the right amount and type of carbon nanotubes. As a result, the inexpensive plastic used to make optical discs will feature in high-end military aircrafts to shield them against build up of electrical charges and pulses which can lead to significant failures.

Additionally, by modifying the amount of carbon nanotubes added to the polycarbonate-nanotube mix, the electrical conductivity of the nanocomposite could be changed from that of silicon to a few orders below what is achieved by metals.

Films

The presence of filler incorporation at nano-levels has also been shown to have significant effects on the transparency and haze characteristics of films. In comparison to conventionally filled polymers, nanoclay incorporation has been shown to significantly enhance transparency and reduce haze. With polyamide based composites, this effect has been shown to be due to modifications in the crystallisation behaviour brought about by the nanoclay particles; sperilitic domain dimensions being considerably smaller. Similarly, nano-modified polymers have been shown, when employed to coat polymeric transparency materials, to enhance both toughness and hardness of these materials without interfering with light transmission characteristics.

Environmental Protection

Water laden atmospheres have long been regarded as one of the most damaging environments, which polymeric materials can encounter. Thus an ability to minimize the extent to which water is absorbed can be a major advantage.

Available data indicate that significant reduction of water absorption in a polymer could be achieved by nanoclay incorporation. Similar effects could also be achieved with polyamide-based nanocomposites.

Specifically, increasing aspect ratio diminishes substantially the amount of water absorbed, thus indicating the beneficial effects likely from nanoparticle incorporation compared to microparticle loading.

Hydrophobicity enhancement would clearly promote both improved nanocomposite properties and diminish the extent to which water would be transmitted through to an underlying substrate. Thus applications in which contact with water or moist environments is likely could clearly benefit from materials incorporating nanoclay particles.

Food Packaging

The gaseous barrier property improvement that can result from incorporation of relatively small quantities of nanoclay materials has been shown to be substantial. Data provided from various sources indicate oxygen transmission rates for polyamideorganoclay composites, which are usually less than half of the unmodified polymer.

Further data reveals the extent to which both the amount of clay incorporated in the polymer, and the aspect ratio of the filler contributes to overall barrier performance. In particular, aspect ratio has been shown to have a major effect, with high ratios (and hence tendencies towards filler incorporation at the nano-level) quite dramatically enhancing gaseous barrier properties.

Development of a combined active/passive oxygen barrier system for polyamide-6 materials is underway at various laboratories across the world. Passive barrier characteristics are provided by nanoclay particles incorporated via melt processing techniques whilst the active contribution comes from an oxygenscavenging ingredient.

Oxygen transmission results reveal substantial benefits provided by nanoclay incorporation in comparison to the base polymer (rates approximately 15-20% of the bulk polymer value, with further benefits provided by the combined active/passive system).

Increased tortuosity provided by the nanoclay particles essentially slows transmission of oxygen through the composite and drives molecules to the active scavenging species resulting in near zero oxygen transmission for a considerable period of time.

Such excellent barrier characteristics have resulted in considerable interest in nanoclay composites in food packaging applications, both flexible and rigid.

Specific examples include packaging for processed meats, cheese, confectionery, cereals and boil-in-thebag foods, also extrusion-coating applications in association with paperboard for fruit juice and dairy products, together with co-extrusion processes for the manufacture of beer and carbonated drinks bottles. The use of nanocomposite packaging would be expected to enhance considerably the shelf life of many types of food.

Fuel Tanks

The ability of nanoclay incorporation to reduce solvent transmission through polymers such as polyamides has been demonstrated. Available data reveals significant reductions in fuel transmission through polyamide–6/66 polymers by incorporation of nanoclay filler.

As a result, considerable interest is now being seen in these materials as both fuel tank and fuel line components for cars. Of further interest for this type of application, the reduced fuel transmission characteristics are accompanied by significant material cost reductions.

Nanocomposites are currently being used in a number of fields and new applications are being continuously developed. Other applications for nanocomposites include:

- Thin-film capacitors for computer chips.
- Solid polymer electrolyes for batteries.
- Automotive engine parts and fuel tanks.
- Impellers and blades.
- Oxygen and gas barriers.

Applications of Nanocomposite Materials in the Oil and Gas Industry

Polymers and their composites, and elastomers have performed well within their operational envelope in gas, oil and petrochemical environments. These materials have unique characteristics that enable them to solve and avoid some of the frequent problems these industries confront. Some operational conditions require these materials to have an enhanced behavior under the combination of chemical, thermal and mechanical parameters. Therefore, the operational envelope of these nonmetallic materials should be broadened.

In the last decade, an important investment has funded the development of products based on nanotechnology, specifically nanocomposites, to increase the performance of polymer matrices without compromising their original features. This development results in very little changes of the final component manufacturing process. A small amount of nanoparticles has a significant influence on improving material performance, with slight changes on processing parameters and using the standard polymer processing equipment.

Type of Nanoreinforcements

There are two main types of layered reinforcements: clay and graphite. Both materials exist as layered materials in their bulk state. The layers must be separated and dispersed throughout the matrix phase, passing from a particle-reinforced composite with an area aspect around 10 for poorly dispersed particles, to high area aspect, between 100-1000, for exfoliated and dispersed nanocomposite. Figure illustrates the morphology of polymer reinforced nanoclays. These can be structurally found in three different types: Layered particles (non-nanocomposite), intercalated and delaminated or exfoliated. The latter nanostructure is necessary to optimize the enhancement of the nanocomposite properties and show greater homogeneity than intercalated nanocomposites.

Enhancement of the thermal and mechanical properties as well as reduction of the rate of water absorption and coefficient of thermal expansion (CTE) in the flow direction, was studied by the work performed by Toyota Research Lab. Just by adding 5% of layered silicate (Montmorillonite - MMT) to polyamide 6, the tensile strength increased by 40%, the tensile modulus by 68%, the flexural strength and modulus by 60% and 126%, respectively; and the heat distortion temperature changed from 65 °C to 152 °C.

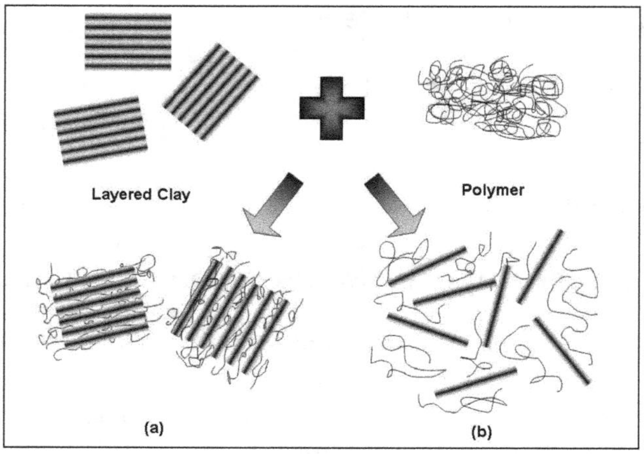

Morphologies of nanocomposite, (a) intercalated and (b) exfoliated.

In addition to improving the thermal-mechanical properties of polymers, nanoreinforcements have been used to enhance the thermal stability and flammability behaviour of nanocomposite materials. Several studies of the thermal stability and flammability of thermoplastic and thermosetting resins, reinforced with nanoparticles, have demonstrated their increase of the temperature of degradation. In addition to the increase of the degradation temperature, the cone calorimeter evaluations on layered silicates (clay) nanocomposites have shown lower peak and average heat release rates (HRR), and longer burning times compared to the control polymer samples.

Gilman et al has demonstrated that the presence of nanoreinforcements improves the flammability behaviour of thermoplastic and thermosetting polymers. Gilman et al suggested the nanocomposite flame-retardant mechanism. A high performance carbonaceous-silicate char builds up on the nanocomposite's surface during burning. This insulates and slows the mass loss rate of decomposition products. This char layer forms as the polymer burns away and the silicate layers re-assemble into the multilayer char. This multilayer char will prevent the gases generated during combustion to escape and be in contact with the flames, diminishing the heat force.

The nanoclays can also be used synergistically in combination with the traditional fire retardant systems and additives to improve the fire performance of polymer materials. Adding nanoclays in combination of the fire retardant components, such as brominated systems, alumina trihydrate (ATH), magnesium hydroxide (MDH), and Ammonium Polyphosphate (APP); will reduce the amount of these harsh compounds in the polymer, balance the mechanical properties, low burn temperature, slow flame spread, reduce specific gravity, enhance processability, reduce blooming, etc.

Other features of the nanocomposite materials reinforced with layered silicates (clays) are the improvement in the resistance against environmental attack and reduction on the gas permeability rates. Ogasawara et al found a lower gas diffusivity rate with nanoclay reinforced composites compared to neat epoxy resins. Ravindran et al studied the effect of distilled water in glassfiber reinforced plastic (GFRP) with and without 5wt% of montmorillonite nanoclays. Scanning electron microscopy (SEM) studies of these samples have demonstrated progressive degradation of both the matrix and glassfiber reinforcement. The moisture degradation on the composite components decreases with the increase of the amount of exfoliated nanoclays in the material. These results have been corroborated by mechanical testing of GFRP specimens. The other type of nanoplatet-reinforcement is the exfoliated graphite sheet, which holds similar thickness as the exfoliated clay. It shows high tensile modulus, tensile strength, thermal conductivity, and low electrical resistivity comparing to clay platelets. Exfoliated graphite nanocomposites possess low electrical resistivity when percolation threshold value of the conductive phase is reached.

Nanostructured chemicals are in the range of 0.7 to 50 nm, which means they are smaller than macromolecules. Particularly, POSS molecules have a hybrid chemical composition, intermediate ($RSiO_{1.5}$) between Silica (SiO_2) and silicones (R_2SiO); and the molecules are physically large compared with singles molecules, ranging from approximately 1-3 nm. They have higher thermal and chemical resistance than silicones, and their nanostructured shape and size provide unique properties by controlling polymer chain motion at the molecular level. POSS molecules are tailorable, which allows them to be compatible with different types of thermoplastic and thermosetting polymers and elastomers. Figure shows the anatomy of a POSS nanostructuredTM chemical. R and X groups can be tailored to be compatible with alkyls, olefins, alcohols, esters, anhydrides, acids, amines, imides, epoxies, thiols, sulfonates, fluoroalkyls, silanols, siloxides and others. Moreover, a metal can be incorporated in the cage structure.

Like nanoclays, functionalized POSS molecules have demonstrated the enhancement of mechanical and thermal properties, and the increase of chemical resistance to water and other solvents. In the particular case of epoxy resin formulations, POSS has demonstrated the ability to raise the glass transition temperature and maintain the modulus at a very high temperature. Figure compares the change in mechanical properties (E' moludus) as a function of the temperature for the epoxy control formulation and the POSS modified epoxy. It is clearly noticed how the modified epoxy maintains the mechanical properties even at temperatures higher than 200 °C. Phenolic resins have shown better fire resistant properties when they are modified with POSS enhanced resorcinols, including superior high temperature stability, excellent char retention, and enhanced insulation performance.

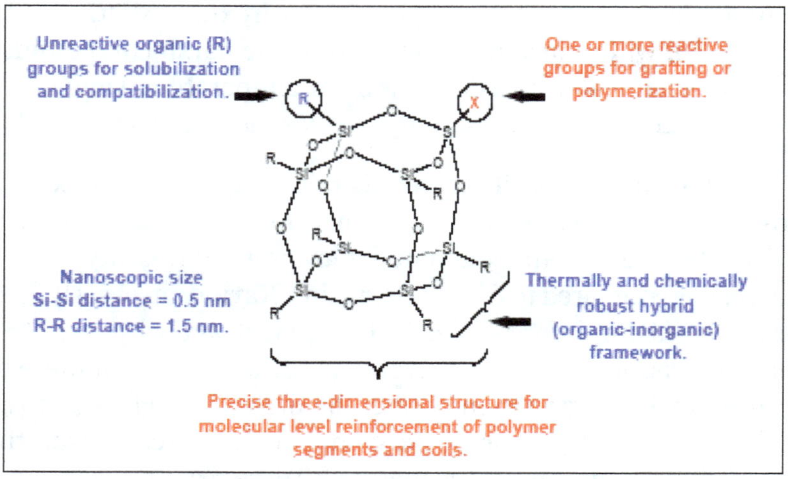

Anatomy of a POSS nanostructured Chemical.

Carbon nanotubes have demonstrated unique mechanical and thermal properties, where the scale of reinforcement was from the micro level in fiber reinforced composites, to a nano level with the carbon nanotubes. The advance in carbon nanotubes has been aimed at the research and development scale. This has been due, as pointed out by Thostenson et al, to the need for development of nanotube production techniques at the scale needed for producing macroscopic composite. In addition to the enhancement of mechanical properties (strength), the carbon nanotubes show low electrical resistivity and good thermal conductivity.

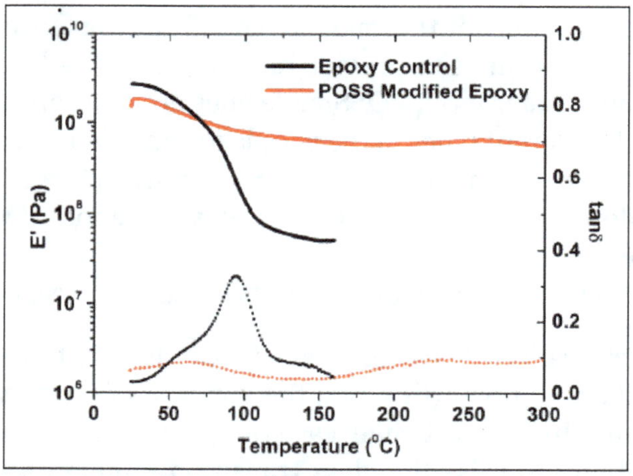

Thermal Properties of epoxy resin reinforced with POSS.

Potential Applications in the Oil and Gas Industry

The use of nonmetallic products in the oil, gas and petrochemical industry has been growing exponentially up in the last two decades. Components and parts made of polymeric and composite materials have been replacing the use of traditional metallic components due to the better corrosion and chemical resistance, lighter weight, lower maintenance costs and other improved features. Some of the typical applications of the nonmetallic materials are fiber reinforced plastic (FRP) pipes, thermoplastic pipes, thermoplastic and composite liners for carbon steel pipes, FRP tanks, composite and thermoplastic valves, composite repair, soft seals and seat for valves and pumps, gasket, elastomeric parts, coatings, etc. The field experience has demonstrated the excellent performance of these nonmetallic materials in the oil, gas and petrochemical industry.

Nevertheless, the industry is demanding nonmetallic materials to meet new challenging conditions, such as higher operating temperatures, stronger and stiffer mechanical properties, improved chemical resistance, lower gas permeability, fire performance, etc. The answer could be found in new materials and technologies, where materials can be tailored to fulfil specific operational and design requirements. Nanocomposite materials offer these potential of assembling materials with different properties to obtain new materials with improved performance.

Thermosetting piping is probably the largest application of nonmetallic materials in the oil, gas and petrochemical industry. Whereas, thermoplastic pipes have their application focused on chemical and gas environments, and as liners for carbon steel pipe in the oil and gas industry. In both types of materials, the need of enhancement of the thermo-mechanical and creep properties will allow to operate at higher pressure and temperature without increasing the component wall thickness. This is very important for high pressure – large diameter pipes, where the pipe wall thickness will have an impact in the cost and installation procedures. Furthermore, thinner pipes will avoid adjusting the fabrication process, e.g., curing process for thermosetting pipes or, processing time to avoid the development of stresses in thick thermoplastic pipe, due to high thermal gradients through the wall thickness.

The chemical resistance to the oil, gas and petrochemical environment at elevated temperature is another feature that can be achieved using nanocomposite material technology. In the case of glass reinforced epoxy pipe, it is known that the epoxy resin can absorb water that will affect the resin's mechanical and creep behaviour. Actually, this factor is considered during the determination of the regression line of the hoop stress in composite pipes. Water absorption effect in the creep behaviour of the nanocomposite material would be diminished, resulting in composite pipe with enhanced long-term properties in service.

Polyethylene can interact with oil components, mainly low weight aliphatic and aromatic components. This physical interaction will allow the hydrocarbon to swell polyethylene material, therefore, a reduction in the mechanical properties of the pipe is observed. With the addition of nanoreinforcement, it is expected that the permeability and absorption of crude oil molecules will decrease. This will reduce the loss in mechanical properties. Moreover, the initial properties of the nanocomposite polyethylene would be higher than the neat polyethylene material, which would also compensate for the properties lost due to fluid absorption. This will be important in the case of polymeric liner, where the material stiffness plays an important role in avoiding liner collapse, mainly in multiphase fluids with the presence of free gas.

Hydrolysis of amide molecules in the presence of water occurs in polyamide material above 70 °C. It is expected the nanoparticles will act as a barrier against water. This will reduce the water absorption rate of the material, and therefore, the hydrolysis effect would be constrained. Therefore, the life cycle of polyamides liner at temperatures above 70 °C would be extended when water is present in the fluid.

To improve the fire performance of composite materials used in the oil and gas industry, some products has been developed using phenolic resin as a matrix base material. Phenolic gratings and polysiloxane-phenolic (PXS) resin pipes are the two main developments which have an excellent endurance performance in the case of fire event. They have good fire resistance, flame spread out control, and low smoke emissions and toxicity. The limitation can be their brittleness, reflected by a low impact resistance compared to other resins. Nanoreinforcement can be used to enhance the impact resistance of the phenolic resin composites, keeping or improving their fire resistance capabilities.

It is well known that the majority of the composite material uses polyester or vinyl ester as the matrix phase to manufacture pipes, tanks, valves, grating, etc. Using nanoreinforcements will be possible to improve the fire resistance behaviour without affecting their mechanical properties and avoiding the use of hazardous materials. The fire resistance properties can be obtained by making the bulk material using nanoreinforcements, or developing intumescent type coating to be applied as an external layer to reduce the effect of fire in the body of the component.

The low resistivity of the carbon base nanoparticles, layered graphite or carbon nanotube, can be used to increase conductivity when the application is required. Per example, for some applications, it is required that offshore fibreglass pipe have the capability to be grounded to avoid the build-up of electrostatic charges. Carbon nanoparticles could be added to fibreglass pipe to impart this electrical feature on the pipe. This same characteristic can be used to develop self-monitoring pipe material. Change in the material resistivity will allow evaluating the effect of fluid absorption and/or mechanical damage, where separation among the nanoparticles would be displayed as an increase in material resistivity.

Application of Nanocomposites in the Automotive Industry

Automotive industry is constantly making efforts to the development and application parts that are light weight, and which at the same time possess excellent mechanical and tribological properties. Today, the automotive industry must meet certain requirements in terms of reducing fuel consumption and at the same time to maintain adequate comfort and safety of the vehicle. Another important factor is the weight of the vehicle, which has a direct impact on fuel consumption and thus the emission of toxic and harmful products of combustion.

The hardest part of the vehicle is the body shell, which makes 40% of its total weight. Therefore, the most common approach to lightweight design integrated within the structure of the vehicle which uses a combination of different materials, depending on their engineering properties and functional characteristics. In recent years increased use of high-strength steel, which in compact European cars makes 50 to 60% of the body. In addition to the use of steel, has increased the use of thermoplastics and thermoset composites used for coating bodywork, bonnet, roof components and other components of the body.

The main materials used in the automotive industry are composites which are made of high-strength steel, aluminum, and composites of carbon fiber and plastic. Applying high-strength steel is achieved by lessening the weight by 20% compared to the conventional steel constructions, use of aluminum of 40% and 50% at the application of composites of carbon fiber. Therefore, by using light materials, reduces the weight of the vehicle, for example, the weight of aluminum ties in the suspension is smaller by 15% than the conventional, hollow composite shaft is 18% lighter than the conventional, and the application of forged aluminum rim saves up to 36% when compared to steel wheels.

Aluminum as a very light material is used as the main matrix element in the production of composite materials referred to as aluminum metal matrix composites (AMCs). Application of aluminum composite is widespread, from the military, automotive, flights until the space industry. AMCs compared with primary aluminum alloy have a good combination of mechanical and tribological properties. In addition, these properties can be tailored to the specific requests. AMCs possess improved physical and mechanical properties such as strength to weight ratio, a good ductility, a high strength and a high modulus of elasticity, low thermal expansion coefficient, good wear resistance, good corrosion resistance, high temperature creep resistance, and better fatigue. The use of lightweight materials for the production of automotive components also contributed to the reduction of air pollution.

Nanocomposites represent a new class of materials that exhibit excellent thermal and mechanical properties. Due to the high performance are suitable for use in aerospace, automobile, chemical and transport industry. The first industrial production of nanocomposites in the automotive industry occurred in 1991 with the production of timing belt covers as part of the engine for Toyota Camry cars.

Most research and development based on nanotechnology is being implemented in the automotive sector. Nanotechnology is applied to parts of the body, chassis and tires, automotive interiors, electrical and electronics, IC engines and drive systems. In recent years, remarkable progress has been made in various industries which motivated researchers to work on new structural materials for better performance in engineering applications.

Tiruvenkadam N. et al have made investigations of the use of the aluminum hybrid nanocomposite for the IC engine cylinder liner (NL) analyzing the performance, combustion and emission characteristics of the engine and others. The aim was to replace the piston cast-steel composite material with the base of Al 6061 and Nano reinforcement zirconium dioxide (ZrO_2), silicon carbide (SiC) and Solid lubricant graphite (Gr) to the size of reinforcement 100 nm, 220 μm and 100 microns, respectively. The mass fraction of each reinforcement amounted to 2.25 wt. %. They concluded that the combination of the selected matrix with reinforcement suitable for NL based on the density, mechanical and thermal properties and of a homogeneous mixture is carried out based on SEM analysis. They also found that the application of NL can achieve greater in–cylinder pressure and higher efficiency. Applying NL is reduced emissions of CO, HC and smoke quantity output, but the increased NOx emission.

Carvalho O. et al have applied a hybrid nanocomposite with an aluminum base for making the piston ring. The material that was used is a hybrid nanocomposite with aluminum base reinforced with SiC particles, and nano MWCNT's of the order of 50 nm, produced by hot–pressing technique in order to evaluate the different fraction of CNTs. The best results have shown abrasion in case of the 6 wt. % CNTs, however, because the best ratio between mechanical and tribological properties of a piston rings are selected from the 2 wt. % of CNTs.

Testing of mechanical and tribological characteristics of the nanocomposite is executive for the production of connecting rods are Ramachandra M. et al. The application of aluminum alloys with 99% of purity and reinforcement n–ZrO_2 25 nm in size were made using the nanocomposite powder metallurgy technique for the production of connecting rods. Testing of wear is made to pin on disc wear testing machine. The study came to the conclusion that the nanocomposite has a higher value than the hardness of the material base, and also reduces wear nanocomposites.

The application of the nanocomposite with a base of aluminum A356-reinforced nano–particles of the CNT is used for making disc brakes and the drum. Sundaram M. and Mahamane U. produced nanocomposite by the method of casting with mixing. Tests have shown that an increase in the mass fraction of nano reinforcement CNT based on the A356 leads to an increase in hardness, while reducing wear, and friction coefficient. Increasing the hardness of the nanocomposite with an increase in the value of reinforcement in the aluminum base. They interrogators aluminum nanocomposite reinforcement with SiC is used for the preparation of camshafts. Stir the nanocomposite is produced casting method with the percentages by weight of reinforcement of SiC of 1 to 4 wt. % and a particle size of 40 nm in the Al base. To improve the mechanical and tribological characteristics of aluminum nanocomposite came Asmatulu R. et al. LM6 alloy strengthened reinforcement nano SiC and Al_2O_3 in the relationship 0.5, 1.0, 1.5 and 2 wt. % of a styrene casting method. Based on the results concluded that as such, the composite can be used in the automotive industry for the production of brake discs.

Summarizes the application of a hybrid nanocomposite with aluminum base Al 356 and Al 7075 in the automotive industry have made Dhanabal S. et al. As reinforcement nano size is used various kinds of materials and by methods of making nanocomposites discussed with stirring casting and pressure casting. The study characterization of nanocomposites came to the result that the piston for making the best show nanocomposites with reinforcement SiC–Al_2O_3, Al_2O_3–MoS_2, TiC, Gr, for the piston ring are nanocomposites reinforced with Al_2O_3–graphite, drum brakes for the reinforcement TiB_2–SiC, the shaft reinforcement is SiC; for camshaft reinforcement SiC–CNT, Al_2O_3–TiC and Al_2O_3 for valves.

Composite Materials

Composite materials are produced by adding basic material of the reinforcement in order to improve certain characteristics, depending on the purpose of the selected base material and the reinforcement. The composites are used as a basic structural material in the aerospace and automotive industry is comprised of deformable metal base and the rigid ceramic as reinforcement. These compositions are known as a composite with a metal base (MMC–Metal Matrix Composite). The first level of classification of composite materials depending on the base material, in addition to existing MMC polymer matrix composites (PMC–Polymer Matrix Composite), a ceramic matrix when it is about composites with ceramic matrix (CMC–Ceramic Matrix Composite).

The second level of classification of composites relates to the shape of reinforcement - the particles, short fibers, laminar composites with continuous fibers and woven composites. In this category are braided and knitted fiber bundles (short fibers, continuous fiber laminated composites) and woven composites (braided and knitted fiber architectures are included in this category).

When the composite material has reinforcement in which at least one dimension in the nanometer scale or less than 100 nm, then the case of nanocomposites. Therefore, the nanocomposites are composites that consist of the base material and the nano size of the reinforcement which do not dissolve in one another.

Nanocomposites reinforced particles are largely used in the automotive industry due to their ability to withstand high temperatures and pressures. Several manufacturing methods used to produce them. The main disadvantage of these materials is a lack of homogeneity in the dispersion of the particles as and weak links between the matrix and the particles. However, research and testing of nanocomposites are still underway about the process of obtaining and variations of certain parameters in order to overcome the aforementioned shortcomings.

Metal Matrix Nanocomposites

Nanocomposites reinforced particles are largely used in the automotive industry due to their ability to withstand high temperatures and pressures. Several manufacturing methods used to produce them. The main disadvantage of these materials is a lack of homogeneity in the dispersion of the particles as and weak links between the matrix and the particles. However, research and testing of nanocomposites are still underway about the process of obtaining and variations of certain parameters in order to overcome the aforementioned shortcomings.

Metal Matrix Nanocomposites (MMNC's) use ductile metallic matrix in which the ceramic reinforcement adds. Various properties of these composites is improved, including the density, the hardness, abrasion, deformation and corrosion resistance depending upon the reinforcement (type, type and content) and the matrix as a manufacturing process. MMNC's become important because composites are used in the industry.

Nano–particles as a reinforcement in metal matrix composites have been progressively replaced other types of reinforcement such as nano–fibers, nano–whiskers or nano–platelets. The main division is reinforcing the oxides, carbides, nitrides and borides. The most commonly used types of nanoparticles are: SiC, TiC, WC, TaC, TiB_2, AlN, and Al_2O_3.

For an easy and high–quality use in automobile components as a template for the production of MMNCs used low density metals such as aluminum (Al), magnesium (Mg), and titanium (Ti). Aluminum and aluminum alloys have found many applications in addition to the automotive and aerospace industry in conjunction with carbon–reinforced composites, and silicon carbide. Use of aluminum alloy (such as Al–Si alloy) for a matrix is preferred because of its advantages, including low cost and easy workability.

The most important production methods that are used for aluminum composites are: stir casting, squeeze casting, compo casting and infiltration, spray deposition and powder metallurgy. Stir casting ranks first among all the available techniques because of the low cost, the possibility of using a wide range of materials as well as on processing conditions, where they can be made with compositions up to 30% in the volume fraction of reinforcement.

Basic classification process for obtaining nanocomposites can be classified into three groups:

- Solid state processes include different powder metallurgy techniques with modifications in the processing steps such as, high–energy ball mill, hot pressing, hot isostatic pressing, cold pressing followed by sintering treatment and extrusion.
- Liquid state processes include different processes such as casting, stir casting and squeeze casting list.
- Semi–solid processes include recasting technique with its variants such as compo casting or in combination with squeeze casting list.

Figure below presents the predictions of application of aluminum in different types of vehicles.

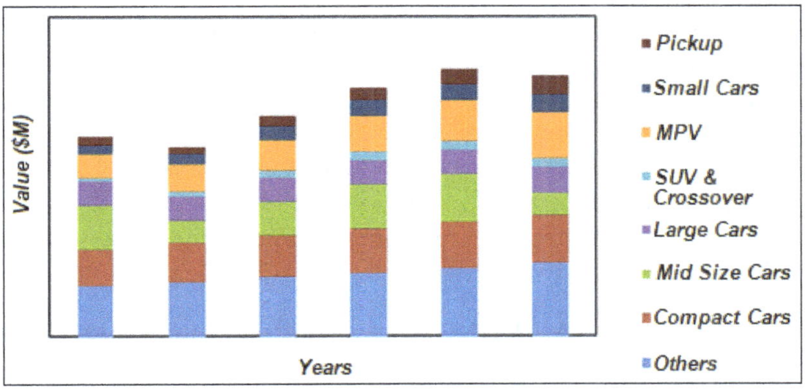

Application of aluminium.

Typical automotive parts which are made of a composite based aluminum alloys as a cylinder head (cylinder head), the pistons, the wheels, crankcase, cylinder head cover, intake manifold, the intake system and the drive means such as fork for a change–speed, drive shaft, gears, gear housing and so on. Application of aluminum castings in the field of automobile industry followed the development of the casting process, the melt treatment and thermal treatment. Great efforts have been made with the aim of improving mechanical and tribological characteristics, by controlling the microstructure that is by improving the processes for preparation, treatment of the metal melt and the heat treatment.

Research of composite materials precisely nanocomposites is still in progress. Researchers have been testing the nanocomposite concluded that they can be applied for the production of the respective parts of cars, such as the piston rings, the pistons, the connecting rods, the disc brakes and the drum brake, the camshaft, the shafts and the valves.

Nanoparticles are common by–products of combustion of oil and other hydrocarbon compounds, combustion and exhaust gases. As stated above, in the automotive industry uses a lot of materials with nanoparticles, and in recent years studies have shown that these nanoparticles can be found in water, air, soil, and therefore in the bodies of animals and humans. These results have led to a debate about toxicity and impact on the environment nanocomposites. Method maturities of nanoparticles in the human body by inhalation, absorption through the skin and if swallowed. Diseases that occur accumulation of nanoparticles in the body of the inflammatory airway disease, bronchitis, asthma, lung cancer, neurodegenerative diseases, cardiovascular disease, liver cancer, Parkinson's disease, Alzheimer's disease, Crohn's disease, and others. Because of these dangers should be given a lot of attention to protecting people during manufacture, transport, handling, use, disposal and recycling of parts and machines which in themselves have nanocomposites.

Polymer Matrix Nanocomposites

The polymers are excellent in terms of low density, ease of machining and the corrosion resistance, however, their rigidity and strength are much less than is required for structural applications. The polymers are considered to be suitable for structural applications matrices, due to their light weight, light processing and long–term stability.

Traditional micro reinforcement shall be replaced by nano–reinforcement in a polymer matrix, making a polymeric nanocomposite. These reinforcement leads to a significant improvement of certain structural and functional features of the polymer in relation to the polymer micro–reinforced reinforcement.

Polymer matrix nanocomposites (PMNCs) are a combination of materials wherein one dimension of the reinforcement is less than 100 nm. The main conditions for the development are the low density PMNCs, ecological and high mechanical properties with a high production speed, good surface finish and with improved corrosion resistance, as well as better operating temperatures. The preparation of nanocomposites uses a wide range of polymers, of which the most commonly used elastomers, thermoplastics and thermosets.

With the development of nanotechnology, various types of nano–reinforcement such as silicate nanoparticles are produced and used in the polymer materials to create so–called nanocomposites of polymer matrix (PMNCs). Uniform dispersion of nanoparticles in the polymer matrix leads to a significant improvement of the properties of the nanocomposites as compared to those of conventional reinforced composites micro particles.

When the polymer matrix is added a small amount of the nanoparticles leads to improve the mechanical properties, the increase in resistance to cracking and propagation, as well as reducing the coefficient of thermal expansion (CTE) of the PMNCs. As reinforcement in polymeric materials for making these PMNCs, silica nanoparticles are used due to relatively high elastic modulus, low density, and good abrasion resistance, low coefficient of thermal expansion, high thermal stability, and low running costs.

There are different processes for the preparation of polymer nanocomposites and each of them has its advantages and disadvantages. As well as in the manufacture of nanocomposite of metal and this must be taken into account for selecting the appropriate method, which corresponds to the target application of the material, as well as the material properties, the ratio of the base and reinforcement, as well as the dispersion of the target performance. Solvent mixing, melt mixing, and in situ polymerization are some of the most common routes for nanocomposite synthesis.

Automotive market has the largest market share of the polymeric nanocomposites thanks to widespread use such as a drive mechanism, the inner and outer power train, suspension and braking systems, exhaust systems and catalytic converters, lubrication, tires and body parts.

On the basis of more stringent legal regulations regarding standards of reducing emissions of cars is expected that the market of polymer nanocomposites recorded a significant increase in applications in the automotive and defense industries, as well as in aerospace engineering. It is expected that the production of nanocomposites result in reducing carbon dioxide emissions from cars. In addition to the automotive industry, the use of polymer composites is increased and the airline industry due to increased production of laminates, sandwich structure, lightning rods and anti-radar protection. Besides the already mentioned application of polymer composites are also used in bio–medicine at the recording of biological, cell tracking, the magnetic separation of biological, nanomedicine.

Other potential applications PMNCs in the automotive industry are as follows: the cap of belt gear unit, the bonnet, fuel hoses, fuel valves, door frames, backrests, fuel hoses, and fuel valves, door frames, seat backs, step assists, heavy duty electrical enclosures, sail panel, box rails, fascia, grills, hood louvers, instrument panels, side trims, body panels, and fenders, sensors, fuel cells. Figure shows the application of polymer nanocomposites with the parts of the car.

The automotive industry invests more than 5% of annual turnover in research and development, and the main focus is on the development of better coatings and paints and stronger, more robust parts for end use. Therefore it is expected that the use of polymeric nanocomposites car market growth over the next ten years.

Overall, the use of plastics in the global market is promising for passenger cars, light commercial vehicles and heavy commercial vehicles. It is anticipated that the global plastics market by 2023 jump to 25.2 billion, and is predicted also a growth of 2.6% in the period from 2018 to 2023.

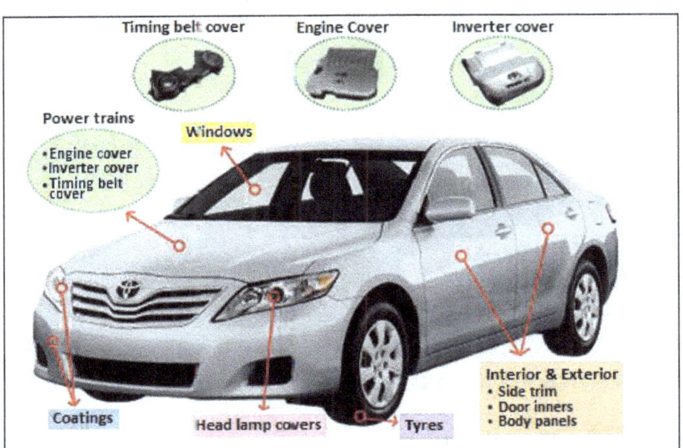

Illustration of the usage of polymer nanocomposites parts.

New trends have directly influence the dynamics of the plastic industry, including the development of technologies for recycling of plastic, and replace PA 66 with cheaper and more advanced PA 6.

The diagrams, figure shows the prediction of application of polymeric materials from 2018 to 2023.

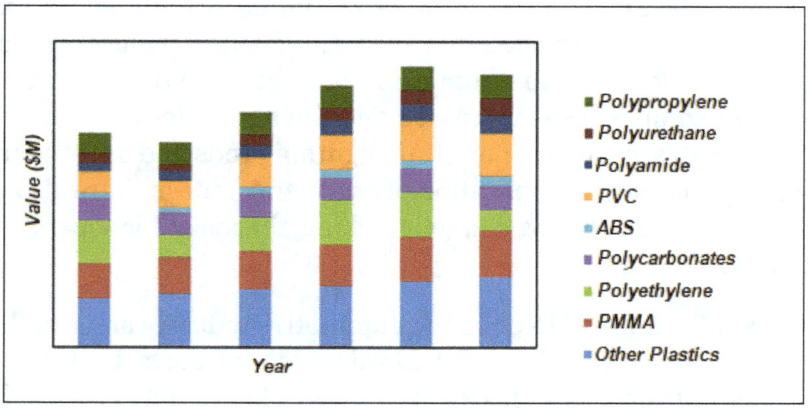

Histogram display forecasting application of polymeric materials in cars from 2018 to 2023.

Trends, opportunities and market forecasts in the automotive industry until 2023, figure by type of vehicle: Passenger cars, light-duty commercial vehicles and heavy-duty commercial vehicles.

Based on the diagram in figure can be concluded that in the next five years, provides for reducing the use of heavy vehicles, indicating an increase in the application of new materials, i.e. nanocomposites.

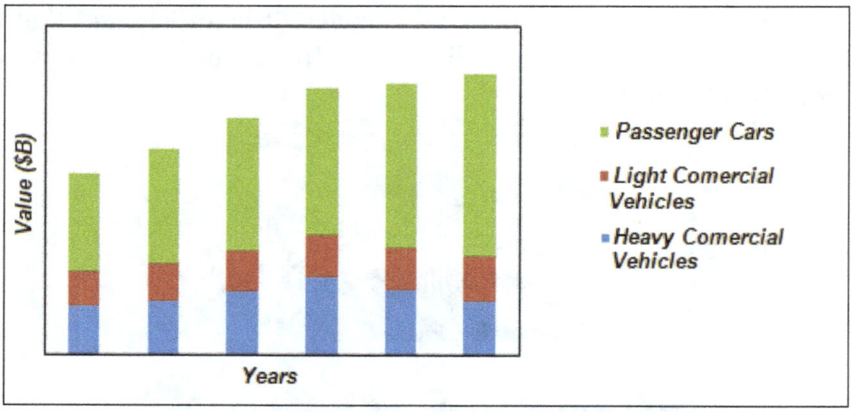

Histogram display forecasts car market in the world.

Ceramic Matrix Nanocomposites

Ceramic composites with dispersed metal particles represent a class of promising material for high performance applications in harsh environments, such as high temperature. These materials offer the possibility of combining a resistance to heat, resistance to degradation, and the abrasion resistance of the ceramic phase having a high mechanical strength and thermal conductivity is provided with metal phase.

Ceramic matrix nanocomposites (CMNCs) is used for making nozzle assemblies, materials stoves, systems for converting energy, gas turbines, thermal engines, etc. The ceramic materials are insensitive to extremes of temperature as opposed to the steel at a high temperature changes the mechanical properties. In particular, they showed good composite materials reinforced with carbon fibers, which are very durable at high temperatures, have a high wear resistance, and have substantially lower specific weight compared with steel and aluminum. Carbon fiber reinforced composites are increasingly being used for car parts that are exposed to high loads, such as e.g. brake disks, valves, cylinder liners, spark plugs, sensors, isolators, filters, piston and others. Safe driving is achieved by applying the brake discs made of ceramic composites, because the discs have a great resistance to deformation and wear, thereby enabling a great savings in weight.

Advantages of using the ceramic composite material in the automotive industry are: The ability to create very complex shapes, reduce the cost of after-treatment of parts, the possibility of connecting the parts during the manufacturing process, dimensional stability in extreme working conditions, corrosion resistance, ease of maintenance, longer life time, possibility to work-up and recyclability.

Disadvantages application of ceramic composite materials in the automotive industry are: The high cost of production, no serial production, the potential for products that are not in use, lack of skilled labor, cruelty (no deformability), sensitivity to moisture and temperature, inability to repair, toxicity, flammability, etc. Another downside application of ceramic materials is the tendency to fracture, expensive process of production and insufficient research.

Compared to conventional CMCs ceramic nanocomposites have improved hardness, high strength, toughness and creep resistance, thermal shock resistance, flame retardancy, magnetic and optical properties, chemical inertness, wear resistance and low density. In addition to automotive applications because of their thermal shock resistance and flame retardancy are used for high-temperature thermal shock zones, such as the fusion energy applications, in gas turbines, aerospace applications, cutting tools, rock drill tips, tools and dies as well as general wear parts, and many others.

The interest of science in recent decades toward the magnetic nanocomposites because of their functional properties for which are widely used starting from the field of magnetic fluid catalysis and adsorption, biotechnology, biomedicine, magnetic resonance imaging, data storage, electronics, magneto-optic and photo catalysis to supercapacitors, hyperthermia, drug shipments, remediation of water, collecting energy. When environmental and biomedical and catalytic apply magnetic metal–ceramic nanocomposites because of its porous structure. These composites are composed of a dispersion of nano–particulate Fe^{o} , CO^{o} , or Ni^{o} nanoparticles usually in amorphous silica or alumina ceramic matrix.

Status of Application of Nanocomposites in the World in the Automotive Industries

Improving the performance of vehicles in terms of quality elements as well as the reduction of greenhouse gas emissions is the constant tendency of the automotive industry. The main driver of the motor vehicle industry is to increase fuel efficiency and reducing vehicle weight.

The application of composites in cars is affected by high production as well as stringent market regulations. In the following chart according to research provides an overview of the car market in the US with revenues by products in recent years and forecasting to 2025, figure.

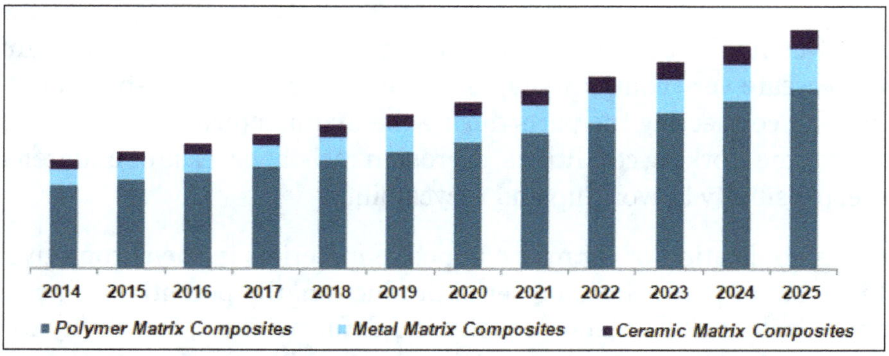

The car market in the US revenue by product, 2014 – 2025 (USD Million).

Growing of carbon emissions due to increased number of vehicles in recent years in the world, launched by governments around the world to react and reconsider the emissions from vehicles. Countries such as China, Japan, Vietnam and Indonesia have implemented various policies in terms of emission standards.

Car manufacturers and composites are well acquainted with the regulations of the countries and producing products that meet specified standards. Automakers believe that reducing vehicle weight is one of the most effective solutions for ensuring compliance and reducing emissions. For every 100 pounds per 45 kg (from 2 to 3%), results in low emissions.

According to studies of the global vehicle market is projected to grow by 5.3% in the period from 2018 to 2023 with great possibilities in the areas of passenger cars, commercial vehicles and heavy commercial vehicles segment. The main growth drivers for this market are increasing production of the vehicle, increasing the need to improve the ride quality and the growing need for advanced systems of safety and comfort.

There is a possibility of growth in a variety of applications, exterior, interior, chassis systems, body and others. It is anticipated that the European market of automotive composites by 2021 will reach $ 4.1 billion and is projected to grow by 5.8% from 2016 to 2021.

The main growth drivers for this market are increasing car production and increasing demand for lightweight materials. The European Union has set new standards require that passenger cars must meet CO_2 emission target of 95 g·km^{-1} (equivalent to 57.9 mpg) by 2021, and for the production of light vehicles is 147 g·km^{-1} (equivalent to 43.3 mpg) by 2020 years.

New trends that influence the dynamics of the industry include increasing penetration of thermoplastic and carbon composites. Another emerging trend is in the formation of strategic alliances between the OEMs and the supplier of carbon fiber and the resin in the automotive industry.

Nanocomposites in Food Packaging Applications

Nanotechnology is the science of engineering materials that have at least one dimension at the nanometer scale (1–100 nanometers). By the use of nanotechnology, we can make structures with new characteristics that are attributable to their nanometer dimensions. One of the fastest growing fields in nanotechnology is nanocomposite food packaging, which is a good alternative for conventional packaging. Usually, composites are composed of a polymer matrix or a continuous phase and a discontinuous phase or filler. Nanocomposites are made by the incorporation of nanofillers that have nano-scale lengths in at least one dimension. In the last few decades, the use of synthetic polymers in food packaging has increased significantly because they are versatile, inexpensive, easy to process, and lightweight. However, plastic polymers are not biodegradable, so they cause environmental problems because it can take hundreds of years for them to be completely degraded. Since these polymers cause environmental pollution, researchers have been trying to develop new packaging materials that degrade more quickly. Biopolymers originate from natural polymers, such as polysaccharides and proteins, and they have attracted widespread attention over the last decades because they are biodegradable. Unfortunately, there are some restrictions on the use of biodegradable or natural polymers in food packaging since they can be poor barriers and have adverse mechanical properties, making them unsuitable for packaging foods. Obviously, in order to increase the shelf life of food, packaging must have low permeability to gases, such as oxygen and water vapor. However, biopolymers are not good barriers to water vapor because of their hydrophilic nature.

Diverse methods can be used to improve the barrier properties of natural polymers, including the use of polymer blends, high-barrier coating materials, and multi-layered films that contain a high-barrier film. In addition to these strategies, a novel technique for this purpose is the use of nanocomposites in food packaging. Nanocomposites improve barrier properties and have a positive impact on the thermal and mechanical characteristics of the packaging material. However, the main concern about applying nanotechnology in food packaging is related to the tiny sizes of the particles of nanomaterial, which can cause them to have different chemical and physical properties than their macroscale chemical counterparts, giving them the potential for causing health problems.

Important Nanoparticles

To date, many nanoparticles have been identified as fillers for making polymer nanocomposites to improve their packaging performances. Among them, clays and silicates have attracted significant attention due to their layered structures. This is because of they are abundant, inexpensive, easy to process, and provide considerable enhancements. There are three main polymer-clay morphologies, i.e., tactoid (or phase separated), intercalated, and exfoliated. In the tactoid structure, which usually occurs in microcomposites, the polymer chains and the clay gallery are immiscible because they have poor affinity for each other. Nanocomposite structures do not display this morphology. In ideal polymer-clay nanocomposites, high affinity would exist between the polymer and clay, leading to exfoliated structures in which polymer chains penetrate into the interlayer space of the clay, making single sheets. If the clay shows a moderate affinity for the polymer, the results would be intercalated structures. There are reports of other particle fillers being used, including silver, zinc oxide, titanium dioxide, carbon nanotubes, graphene nanoplates, copper, and copper oxides. It is studied that graphene nanoplates (GNPs) are able to form nanocomposites with improved heat resistance and barrier properties, making them a great option for food-packaging applications. Carbon nanotubes (CNTs) are another type of carbon-based nanoparticles that have good electrical and mechanical characteristics. However, their use has been hindered, mainly due to their high cost and the difficulty in processing dispersions. Copper, copper oxide, titanium dioxide, zinc oxide, and silver have been used mainly for their anti-microbial properties.

Types and Characteristics of Nanofillers

Type of Nanofiller	Morphology/Structure	Typical dimensions*	Density
Montmorillonite (MMT)	Platelets	D: 100–500 nm, T:~ 1nm [a]	~ 2.6 g cm^{-3}
Layered double hydroxide (LDH)	Platelets with distinct hexagonal shape	D: 50-few hundred nm, T: ~0.5 nm[a]	1.5 g cm^{-3}
Carbon nanotubes (CNT)	Tubular, seamless cylinder of graphene sheet	L: up to tens of μm, D: ~ 1/5/20 nm[b]	1.3–1.5/1.5/1.8–2.0 g cm^{-3}[b]
Cellulose nanowhiskers (CNW)	Rod-like	L: 100 nm-few μm, D: 2–10 nm	~ 1.6 g cm^{-3}
Microfibrillated cellulose (MFC)	Highly fibrous network consisting of bundles of nanofibrils	L: few μm, D: 10–100 nm	~ 1.6 g cm^{-3}
Bacterial cellulose (BS)	Ribbon-shaped nanofibers[c] forming a highly fibrous network	L: few μm, W: 50–150 nm, T: 5–10 nm	~ 1.6 g cm^{-3}
Chitin whiskers (CHW)	Slender rods	L: 50–650 nm[d], D: 4–80 nm[d]	~ 1.5 g cm^{-3}
Starch nanocrystals (SNC)	Platelet-like Nanoparticles; often forming few μm size aggregates	L: 20–150 nm[d], W: 15–30 nm, T 6–8 nm	~ 1.55 g cm^{-3}

Nanocomposites as Degradable Improved Packaging

Currently, there is significant interest in developing different packaging materials because of the increasing demand for foods with minimum processing and longer shelf life. Obviously, such new packaging materials must have excellent barrier properties to prevent the migration of oxygen, carbon dioxide, water vapor, and flavor compounds. This will have a major influence on the shelf life of fresh and processed foods. In the case of using biopolymers, lowering water vapor permeability is especially important due to their hydrophilic nature. However, by the aid of nanocomposite technology, this inherent defect of biopolymer-based packaging materials can be conquered. In many cases, it has been studied that the barrier properties can be improved by about 50% compared to the properties of the neat polymer. This is because of the creation of a maze structure results in a tortuous path for gases and other molecules, thereby reducing their permeation rate. Studies have indicated that incorporating nanocomposites in food packaging materials results in better mechanical and thermal behavior of the packaging. For instance, researchers have studied that engineered nanocomposites of biopolymer-layered silicate have noticeably enhanced physical properties, such as higher tensile strength, enhanced thermal stability, and better gas barrier properties. By using nanocomposites, the food packaging can better tolerate thermal stress of food processing, shipping, and storage. Although traditional composite structures may have large amounts of filler (approximately 60% by volume), nanocomposites usually represent significant changes in properties at quite low loads, i.e., less than 2% volume. It is also suggested that, for achieving the best effects of the final nanocomposites, around 5 wt% of fillers is generally desirable. There is adequate evidence proving other benefits of nanocomposite food packaging, e.g., improved stability of sensory properties, such as flavor, better maintenance of color and texture, increased product stability through the food chain, and less spoilage. Biopolymers can be derived from plant materials as well as animal and microbial products, such as polyhydroxybutyrate. Among these biopolymers, starch, cellulose, and their derivatives, as well as proteins, have been used extensively by many researchers to make bio-based nanocomposites. The most-studied biodegradable nanocomposite is starch-clay, which has been investigated for several applications including food packaging.

Nanocomposites in Active Packaging

Some nanofillers, such as silver, zinc oxide, and magnesium oxide, have antimicrobial or antioxidant activities. Incorporation of these nanofillers in polymer or biopolymer matrices leads to an inhibiting or retarding effect on the growth of microorganisms, thereby reducing food spoilage. The main goal of active packaging systems is extending the product's shelf life. They can also be designed to improve food quality and safety and finally result in less food waste. Anti-microbial nanocomposite films are worthwhile because of their anti-microbial properties caused by natural anti-microbial agents and because of their suitable structural integrity, which results from the barrier properties created by the nanocomposite matrix. Nanoscale materials have a higher surface-to-volume ratio than their microscale counterparts, and, therefore, they are able to attach to a vast number of biological molecules, which enhances their efficiency. Previous reports have identified potential applications of nanocomposites as growth inhibitors, bacteriocides, and antibiotic carriers. In addition to the application of nanocomposites as packaging materials, they also can be used as delivery systems by helping the migration of functional additives, such as minerals, probiotics, and vitamins into the food. These controlled-released packagings are another example of a nanomaterial application in active packaging. Many anti-microbial nanocomposites used for food packaging are made from silver, which has an intense toxicity to a large variety of microorganisms. Active nanocomposites are advanced alternatives to conventional active plastic technologies or active sachets for the extension of the quality and safety of packaged food products. The term 'active nanocomposite' generally refers to a plastic composite (i.e., a polymer blend) that contains an active, nanostructured material that confers an activity on the plastic matrix. At least one of the dimensions of the active nanostructured material must be less than 100 nm in size. Nanoclays can be used as carriers for the active agent. The efficacy of the active agent is enhanced because it is highly dispersed in the polymeric matrix and, hence, exposed more efficiently to the substance on which it is required to act.

Nanocomposites in Intelligent Packaging

Intelligent packagings are able to communicate with the consumers and give information about the product condition through the food chain. These packagings can monitor, trace, or record outer or inner changes that are occurring in the product or its environment. By applying reactive components in the form of nanoparticles and making so-called nanosensors, the packaging will be able to respond to environmental changes, such as temperature, the presence of oxygen, deteriorated products, and microbial contamination. Through the incorporation of nanosensors into food packaging, detection of certain chemical compounds, pathogens, and toxins in food would be possible. This also fulfills the need for inexact expiration dates, which, in many cases, are not suitable for the products due to false estimation of product condition during storage. So they provide real-time status of the food's freshness. These sensors have many advantages over the costly and time-consuming conventional detection methods, such as High-performance liquid chromatography (HPLC). This includes their speed, high-throughput detection, ease of operation, cost effectiveness, and decreased power requirements.

Analyzing Techniques of Nanomaterials

Discussing the potential risks of nanoparticles for consumers requires thorough insight about their characteristics. For the characterization of nanocomposite structures, aspects that are more likely to be considered are particle dispersion, changes in the mass matrix, and the type of particle–polymer interface. Detecting the migration of nanomaterial into food matrices requires more sensitive analytical techniques due to the complexity of the nanomaterials and their representing only a very small portion of the bulk food. Detection of this material in complex matrices, such as food, requires combined methodologies, since no individual technique can give all the detailed information. Microscopic, spectroscopic, and quantitative analytical techniques are all useful for this purpose. Transmission electron microscopy (TEM), scanning electron microscopy (SEM), and atomic force microscopy (AFM) are good examples of microscopic methods, each of which has its own advantages and disadvantages. By these methods, nanoparticles' general features, such as size, shape, structure, dispersion, and coagulation state, are accessible. However, these techniques usually are destructive, meaning that verification tests for a better analysis are not feasible on the same sample. Spectroscopic techniques, such as X-ray diffraction (XRD) and UV-vis spectroscopy, can be used for characterization and analysis of nanomaterial as well. XRD, both wide angle (WAXS) and small angle (SAXS), are the conventional methods that are used to examine nanocomposite structures. Elemental composition or the crystalline arrangement of nanomaterials, both in natural and engineered form, can be investigated by the XRD technique. Despite microscopic methods, this technique is not destructive, and there is no need for difficult sample pretreatment.

These attributes contribute to its extensive application in the characterization of materials. Different nanocomposite formations, such as intercalation, exfoliation, and dispersion, conventionally have been studied by XRD. The XRD technique traditionally has been used especially for characterizing montmorillonite exfoliation in polymer materials used in packaging applications. The supporting method, UV-vis spectroscopy, has been used extensively due to its cost effectiveness and ease of operation. This technique confers information about the presence of nanoparticles and their characterization. Nanomaterial quantitation in food packaging applications also is a vital issue because of safety concerns. For this purpose, migrated nanomaterial can be probed through inductively coupled plasma mass spectrometry (ICP-MS), inductively coupled plasma atomic emission spectrometry (ICP-AES), and inductively coupled plasma optical emission spectrometry (ICP-OES), which are three important techniques for quantification and elemental analyses. The advantages of these methods include their high selectivity, sensitivity, and accuracy. Using these methods, nanoparticles can be detected in amounts as low as 0.1–10 ppm. Introducing nanomaterial into the ICP, a beam of gaseous ions is produced by the atoms of the analyte in the plasma, resulting in an individual pulse measured by the detector, which appears as a peak on the graph. Prior to injecting the sample, a digestion process is required because the presence of food components could impede good atomization of the sample. An alternative approach to ICP is atomic absorption spectrometry (AAS) for the detection of nanomaterial. In many cases, AAS is used for the quantitative determination of chemical elements within samples, such as food, by measuring the absorption of optical radiation (light) by free atoms in the gaseous state. However, analysis of more than one element is not possible by AAS.

Nanomaterial Migration into Food Matrix

One of the critical issues in food packaging is migration. Migration is the unintentional transfer of packaging materials into the food. This problem may influence the food's safety and, subsequently, consumers' health. It also can cause undesirable organoleptic changes in the food that is in contact with the packaging. For example, migration of TiO_2 into lipid matrix results in rancidity. Nevertheless, in some active packaging, nanomaterials are intended to be released deliberately. Performing migration tests under controlled conditions is essential for evaluating the possibility of the nanomaterials' incorporation into food packaging and also assessing the associated safety problems. The analysis condition depends on the type of food matrix or food simulant being tested. Different factors impact the migration of nanomaterials into foodstuffs. These factors can be ascribed to the nanoparticles' properties (e.g., concentration, particle size, molecular weight, solubility, and diffusivity in the polymer), environmental conditions (temperature, mechanical stress), food condition (pH value, composition), packaging characteristics (polymer structure and viscosity), and contact time. For instance, higher temperatures and lower pH values increase the solubility of metal nanomaterials in aqueous solution, leading to an increase in their migration. Low molecular weight polymers that have more free volume accelerate the migration rate and nanoparticles' diffusivity. There also is an inverse relationship between the migration rate of a system and the size of the nanoparticles. If the nature of the food is compatible with the type of packaging, the food itself may be absorbed into the polymer matrix, enlarging the gaps between the polymer chains, thereby increasing the migration rate. For example, fats have high affinity for polyethylene (PE) and polypropylene (PP), so they may be absorbed by the packaging and cause an increase in plastic mobility and higher migration rates. The preservation method of foods also is important and can affect nanoparticles' migration. It has been studied that microwave heating causes structural modification of the packaging, thereby speeding up the migration of silver ions. High surface-to-volume ratios in nanomaterials induce an active surface chemistry followed by probably unwanted chemical reactions. This would be troublesome if their presence in the packaging enhances the formation of unforeseen reaction by-products during processing. Nanoparticles also can impact the migration of other packaging constituents, either speeding up or slowing down their migration. To perform migration analysis, the best approach is to use actual food matrices. However, it would be very tedious and difficult to directly estimate the amount of migration into real food matrices due to complex composition of foodstuffs. As a solution to this problem, using natural food stimulants has become an alternative method to measure the specific and overall migration of different packaging substances into foods. The amount and rate of mass transport in the selected food stimulants should be similar to those that occur in the food matrix.

Concerns about Nanocomposite Packaging

The most challenging concern about the use of nanoparticles pertains to their reduced particle size, because the chemical and physical characteristics of such tiny materials may be quite different from their macro scale counterparts. This implies that issues concerning their toxicity may be different based on their very small sizes. This is a very serious problem, because GRAS (generally recognized as safe) decisions probably have been made without any consideration of the size or the related safety evaluations. It has been studied that some nanomaterials are potentially harmful to human health. However, there is still insufficient proof concerning whether this is attributable to any kind of nanomaterial applications, especially in the food industry. Before confirming the safety of nanomaterials, it has been recommended by the Institute of Food Science and Technology (IFST) that nanomaterial be treated as potentially dangerous materials in order to lessen their possible adverse effects on human health. One of the responsibilities of the Food and Drug Administration (FDA) is to prove the safety of nanoscale products. The FDA foundation, for nano-based products are being assessed case-by-case according approved standard. Nevertheless, since there is no labeling prerequisites for such products that state whether nanotechnology has been used in their production, it would be difficult for the FDA to evaluate their safety. Present legislation considers that there is no difference between so-called nano-produced materials and those made by ordinary manufacturing approaches. As a result, nano-based material for packaging are not evaluated as new chemicals, and, thus, specific or overall migration limits into food that is in contact with nanomaterials have not been defined. Regulating principles about migration limits of different plastics into food matrices are applicable for nanoparticles. Even so, this would not fulfill the increasing for food safety due to differences in the physicochemical characteristics of nano and macro-scale components, which cause them to exhibit different safe limits for their migrations. Jen-Yi Huang studied that, currently, there are no internationally agreed-to research protocols or standards concerning nanomaterials.

Polymer Nanocomposites 2

- Polymer Layered Silicate Nanocomposites
- Carbon Nanotube-based Polymer Composites
- Applications of Polymer Nanocomposites

The polymers that have nanoparticles and nanofillers dispersed in the polymer matrix are called polymer nanocomposites. It improves the performance properties of the polymer and has various biomedical and electrical applications. This chapter closely examines polymer nanocomposites to provide an extensive understanding of the subject.

Polymer nanocomposites are materials in which nanoscopic inorganic particles, typically 10-100 A in at least one dimension, are dispersed in an organic polymer matrix in order to dramatically improve the performance properties of the polymer. Systems in which the inorganic particles are the individual layers of a lamellar compound - most typically a smectite clay or nanocomposites of a polymer (such as nylon) embedded among layers of silicates - exhibit dramatically altered physical properties relative to the pristine polymer. For instance, the layer orientation, polymer-silicate nanocomposites exhibit stiffness, strength and dimensional stability in two dimensions (rather than one). Due to nanometer length scale which minimizes scattering oflight, nanocomposites are usually transparent.

Polymer nanocomposites represent a new alternative to conventionally filled polymers. Because of their nanometer sizes, filler dispersion nanocomposites exhibit markedly improved properties when compared to the pure polymers or their traditional composites. These include increased modulus and strength, outstanding barrier properties, improved solvent and heat resistance and decreased flammability.

Polymers that contain transition metal complexes either attached to or directly in a π -conjugated backbone are an exciting and a promising class of modern materials. These macromolecules are hybrid of π -conjugated organic and transition metalcontaining polymers. π -conjugated organic polymers, such as polyacetylene, poly thiophene, and polypyrrole, as well as oligomers and derivatives of these materials have been extensively explored. These materials are endowed with many important properties such as nonlinear optical properties, electronic conductivity and luminescence, and have been proposed for their use in various applications including chemical sensors, electroluminescent devices, electro catalysis, batteries, smart windows and memory devices.

Layered silicate/polymer nanocomposites exhibit superior mechanical characteristics (e.g. 40% increase of room temperature tensile strength), heat resistance (e.g. 100% increase in the heat distortion temperature) and chemical resistance (e.g. ~ 10 fold decrease in o_2 and H_2o permeability) compared to the neat or traditionally filled resins. These property improvements result from only a 0.1-10 vol. % addition of the dispersed nanophase. Polyimide-clay hybrids represent another example of polymer nanocomposites. These nanocomposites have been prepared by intercalation of the organo-clay with a polyamic acid. The claypolyimide hybrid composite films exhibit greatly improved CO_2 barrier properties at low clay content; less than 8.0 vol. % clay results in almost a ten-fold decrease in permeability. Adding nanoscale ceramic powders to commercial products can produce another class of polymer nanocomposites. The addition of reinforcing agents is widely used in the production of commodities (packaging films and tyres). It is expected that the reduction of the added particle size down to nanometric scale could enhance the performance of these materials, even though not to the extent as layer addition. These new materials are aimed at being a substitute for more expensive technical parts (gear systems in wood drilling machines, wear resistance materials) and in the production of barrier plastic film for food industry. Besides structural applications, polymer nanoparticle compounds have very interesting functional applications. For instance, γ Fe_2zO_3/ polymer nanocomposites are used as advanced toner materials for high quality colour copiers and printers and as contrast agents in NMR analysis, memory devices. The key to forming such novel materials is understanding and manipulating the guest-host chemistry occurring between the polymer and the layered compounds or the nanoparticles, in order to obtain a homogenous dispersion and a good contact between polymer and added particle surfaces. There have been major advances in solid state and materials chemistry in the last two decades and the subject is growing rapidly.

The coatings of magnetic particles are of special interest because of their important applications viz. technological energy transformation, magnetic recording, magnetic fluids and magnetic refrigeration system. Polymer materials have been filled with several inorganic compounds in order to increase properties like heat resistance, mechanical strength and impact resistance and to decrease other properties like electrical conductivity, dielectric constant thereby increasing the permeability for gases like oxygen and water vapor.

In recent years considerable efforts have been devoted to the development of methods for the preparation of composite particles consisting of polymer cores covered with shells of different chemical composition. In several of these powders, particles covered with magnetic materials have been used as beads for gas separation, or as pigments, catalysts, coatings, flocculents, toners, raw materials recovery, drug delivery and anti-corrosion protection.

Polymer composites containing ferrites are increasingly replacing conventional ceramic magnetic materials because of their mouldability and reduction in cost. They are also potential materials for microwave absorbers, sensors and other aerospace applications. These flexible magnets or rubber ferrite composites are possible by the incorporation of magnetic powders in various elastomer matrices. This modifies the physical properties of the polymer matrix considerably.

Solvent casting method is one of the easiest methods for the preparation of polymer nanocomposites. It needs simple equipment and is less time consuming. Figures show the optical micrograph images of γ-$Fe_2\ o_3$ dispersed natural rubber ($\gamma Fe_2\ o_3$- NR), γ Fe_2o_3 dispersed polyethylene glycol (γ-$Fe_2\ o_3$ -PEG) and Fe_2o_3 dispersed polycarbonate(γFe_2o_3 -PC), nanocomposite films. The fine dispersion of the magnetite inside the polymer matrix makes it a magnetic polymer. Figure led shows the structure of the polymer polyisoprene (natural rubber).

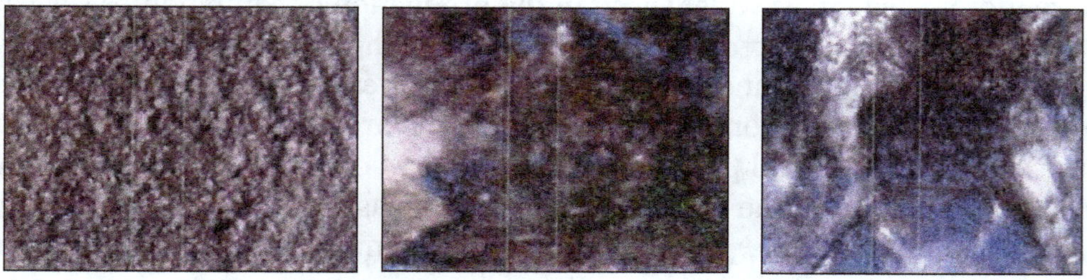

Optical micrograph images of γ -$Fe_2\ o_3$ dispersed (a) natural rubber. (b) polyethylene glycol (c) polycarbonate composite films.

Characterization

The vigorous development of polymeric science and extensive utilization of polymeric materials in technology has led in recent years to the increased interest in the preparation and characterization of polymer and its composite films. Characterization is an essential part of all investigations dealing with materials. The important aspects of characterization are chemical composition and compositional homogeneity (chemical homogeneity), structure (including crystal system where possible atomic coordinates, bonding and ultra structure) and identification and analysis of defects and impurities influencing the properties of the materials. Characterization, therefore, describes all those features of composition and structure of a material that would suffice for reproducing the material. The advances made in the last few years in characterization techniques, especially in the structure elucidation, have been stupendous and have opened new vistas in solid state materials. Among the several characterization techniques, X-ray diffraction (XRD), scanning electron micrography (SEM) and infrared (IR) spectroscopy are the three important techniques.

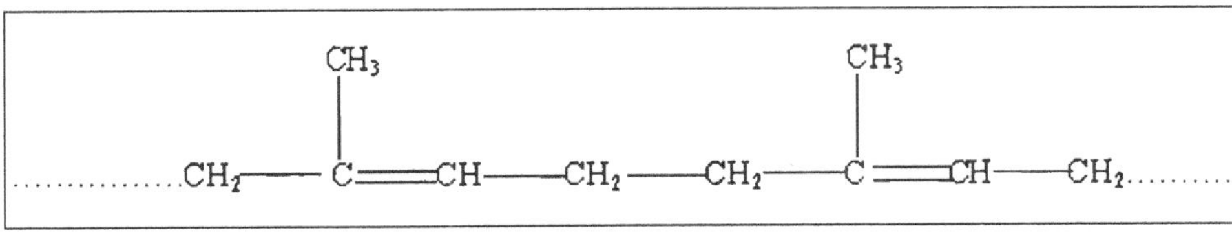

Structure of natural rubber (polyisoprene).

X-ray Diffraction

X-ray diffraction has played a central role in identifying and nature of bonding and the working criteria for distinguishing between short-range and long-range order of crystalline arrangements from the amorphous substances are largely derived from X-ray diffraction and thus it remains as a useful tool to obtain structural information.

X-ray diffraction pattern of amorphous polymer will not show any sharp and highly intensed peaks whereas the nanocomposites of amorphous polymer show sharp and highly intensed peaks. This is due to the development of crystallinity in the amorphous polymer. XRD pattern of pure natural rubber and $\gamma\ Fe_2O_3$ dispersed natural rubber composite. Highly intensed peaks occur in the pattern due to the presence of inserted gamma iron oxide materials in the rubber matrix. X-ray diffraction has been most commonly used for routine characterization as well as for detailed structural elucidation. In order to obtain detailed structural information, knowledge of X-ray diffraction intensities is also essential, the intensities being related to the structure factor.

Scanning Electron Micrograph

Structural phenomena play an important role in determining the properties of a polymer. Mechanical properties are determined not only by the changes in shape confirmation and by motion of individual molecules of the polymers, but by the behavior of larger and more complex structural formations as well. The interface boundaries of these formations, known as super molecular structures, are the sites where chemical reactions in the polymer are most likely to begin and centers of crack formation and incipient destruction are likely to arise. It has been found that extensive occurrences of ordered structures are typical not only of crystalline, but also of amorphous polymers. Despite the complex morphology of structural formation in polymers it should not be forgotten that all these structures are built up of separate polymeric molecules. At a glance it seems self-evident that direct relations must exist between the properties of macromolecules and their ability to form super molecular structures. The shapes of most polymer molecules may vary within wide limits when studying the simplest phenomenon of structure formation. Quite a long time ago it was found that there are two ways by which structures can form. Sufficiently flexible molecules roll up into spherical coil globules, which form in very much the same way as the drops of a liquid under the action of surface tension. But if the macromolecules are sufficiently rigid, the simplest linear structures result. No separate linear polymer molecules have been observed so far. Evidently in majority of cases, they aggregate into chain bunches usually containing several dozen molecules. The phenomenon of structural transformations occurring during deformation is very typical of polymers. A classic example of structural transformation is the formation of a 'neck' on deformation, described for the case of crystalline polymers some time ago by Kakina. It is also observed in the case of amorphous polymers with developed structures and it is firmly described as a phase transformation. The nature of this phenomenon remained obscure for a long time, but electron microscopy revealed that 'neck' formation is actually a jump wise transition from one super molecular structure to another with a sharp interfacial boundary which is also observed on a microscopic scale.

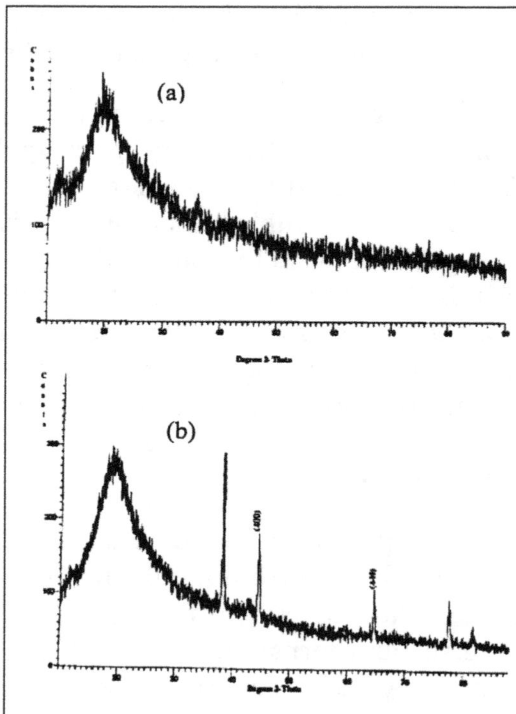

XRD pattern of (a) pure natural rubber (b) γ-Fe_2O_3 dispersed natural rubber.

A well-known example is that of poly(methyl methacrylate) where a sharp boundary can be seen between the isotropic and the oriented parts of the specimen. The formation of a 'neck' on deformation of a large spherulite of isotactic poly (methyl methacrylate) takes place and a sharp boundary can be discerned between the unchanged and the oriented portions of a spherulite. In addition, secondary formations can be seen which have resulted from recrystallisation of the oriented parts, and these are also separated by sharp boundary lines.

Figure shows the SEM images of γ-Fe_2O_3 dispersed natural rubber composite. From the figure one can observe the fine dispersion of iron oxide particles in the rubber matrix. The dispersed particles have irregular shape and show agglomeration.

Infrared Spectroscopy

Infrared spectroscopy is one of the most powerful analytical techniques, which offers the possibility of chemical identification. This technique when coupled with intensity measurements may be used for quantitative analysis. One of the important advantages of infrared spectroscopy over' the other usual methods of structural analysis (X-ray diffraction, electron spin resonance, etc.) is that it provides information about the structure of a molecule quickly, without tiresome evaluation methods. This method can solve many problems in organic chemistry (polymeric materials) and coordination chemistry, and also advantageously complements the results obtained by other methods. This technique is based upon the simple fact that a chemical substance shows marked selective absorption in the infrared region giving rise to dose-packed absorption bands called an IR absorption spectrum, over a wide wavelength range. Various bands will be present in the IR spectrum, which will correspond to the characteristic functional groups and bonds present in a chemical substance. Thus an IR spectrum of a chemical substance is a fingerprint for its identification. IR spectrum of polymer nanocomposite shows the presence of both nanomaterials and polymers (depending upon the polymer chain) at various frequencies.

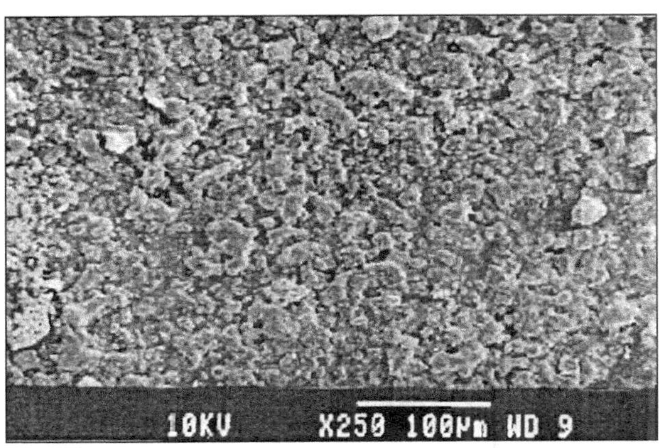

SEM image of γ-Fe_2O_3 dispersed natural rubber.

Thermal Analysis

Thermal analysis may be defined as the measurement of physical and chemical properties of materials as a function of temperature. The two main thermal analysis techniques are thermogravimetric analysis (TGA) which automatically records the change in weight of a sample as a function of either temperature or time, and the differential thermal analysis (DT A), which measures the difference in temperature, ΔT, between a sample and an inert reference material as a function of temperature; DT A therefore detects change in heat content. A technique closely related but modified to DTA is differential scanning calorimetry (DSC). In DSC, the equipment is designed to allow a quantitative measure of the enthalpy changes, (ΔH), that occurs in a sample as a function of either temperature or time.

DSC is an analytical tool which helps to understand the thermal behavior of polymer nanocomposites. It helps in finding glass transition temperature (T_g) of polymer and its polymer composites. The increase in T_g values shows the presence of inorganic materials in the polymer matrix.

Bio-hybrid Polymer Nanofibers

Many technical applications of biological objects like proteins, viruses or bacteria such as chromatography, optical information technology, sensorics, catalysis and drug delivery require their immobilization. Carbon nanotubes, gold particles and synthetic polymers are used for this purpose. This immobilization has been achieved predominantly by adsorption or by chemical binding and to a lesser extent by incorporating these objects as guests in host matrices. In the guest host systems, an ideal method for the immobilization of biological objects and their integration into hierarchical architectures should be structured on a nanoscale to facilitate the interactions of biological nano-objects with their environment. Due to the large number of natural or synthetic polymers available and the advanced techniques developed to process such systems to nanofibres, rods, tubes etc. make polymers a good platform for the immobilization of biological objects.

Bio-hybrid Nanofibres by Electrospinning

Polymer fibers are, in general, produced on a technical scale by extrusion, i.e., a polymer melt or a polymer solution is pumped through cylindrical dies and spun/drawn by a take-up device. The resulting fibers have diameters typically on the 10-µm scale or above. To come down in diameter into the range of several hundreds of nanometers or even down to a few nanometers, Electrospinning is today still the leading polymer processing technique available. A strong electric field of the order of 103 V/cm is applied to the polymer solution droplets emerging from a cylindrical die. The electric charges, which are accumulated on the surface of the droplet, cause droplet deformation along the field direction, even though the surface tension counteracts droplet evolution. In supercritical electric fields, the field strength overbears the surface tension and a fluid jet emanates from the droplet tip. The jet is accelerated towards the counter electrode. During this transport phase, the jet is subjected to strong electrically driven circular bending motions that cause a strong elongation and thinning of the jet, a solvent evaporation until, finally, the solid nanofibre is deposited on the counter electrode.

Bio-hybrid Polymer Nanotubes by Wetting

Electro spinning, co-electrospinning, and the template methods based on nanofibres yield nano-objects which are, in principle, infinitively long. For a broad range of applications including catalysis, tissue engineering, and surface modification of implants this infinite length is an advantage. But in some applications like inhalation therapy or systemic drug delivery, a well-defined length is required. The template method to be described in the following has the advantage such that it allows the preparation of nanotubes and nanorods with very high precision. The method is based on the use of well defined porous templates, such as porous aluminum or silicon.

The basic concept of this method is to exploit wetting processes. A polymer melt or solution is brought into contact with the pores located in materials characterized by high energy surfaces such as aluminum or silicon. Wetting sets in and covers the walls of the pores with a thin film with a thickness of the order of a few tens of nanometers.

Gravity does not play a role, as it is obvious from the fact that wetting takes place independent of the orientation of the pores relative to the direction of gravity. The exact process is still not understood theoretically in detail but its known from experiments that low molar mass systems tend to fill the pores completely, whereas polymers of sufficient chain length just cover the walls. This process happens typically within a minute for temperatures about 50 K above the melting temperature or glass transition temperature, even for highly viscous polymers, such as, for instance, polytetrafluoroethylene, and this holds even for pores with an aspect ratio as large as 10,000. The complete filling, on the other hand, takes days. To obtain nanotubes, the polymer/template system is cooled down to room temperature or the solvent is evaporated, yielding pores covered with solid layers. The resulting tubes can be removed by mechanical forces for tubes up to 10 μm in length, i.e., by just drawing them out from the pores or by selectively dissolving the template. The diameter of the nanotubes, the distribution of the diameter, the homogeneity along the tubes, and the lengths can be controlled.

Applications

The nanofibres, hollow nanofibres, core–shell nanofibres, and nanorods or nanotubes produced have a great potential for a broad range of applications including homogeneous and heterogeneous catalysis, sensorics, filter applications, and optoelectronics. Here we will just consider a limited set of applications related to life science.

Tissue Engineering

This is mainly concerned with the replacement of tissues which have been destroyed by sickness or accidents or other artificial means. The examples are skin, bone, cartilage, blood vessels and may be even organs. This technique involves providing a scaffold on which cells are added and the scaffold should provide favorable conditions for the growth of the same. Nanofibres have been found to provide very good conditions for the growth of such cells, one of the reasons being that fibrillar structures can be found on many tissues which allow the cells to attach strongly to the fibers and grow along them as shown.

Nanoparticles such as graphene, carbon nanotubes, molybdenum disulfide and tungsten disulfide are being used as reinforcing agents to fabricate mechanically strong biodegradable polymeric nanocomposites for bone tissue engineering applications.

The addition of these nanoparticles in the polymer matrix at low concentrations (~0.2 weight %) leads to significant improvements in the compressive and flexural mechanical properties of polymeric nanocomposites. Potentially, these nanocomposites may be used to create novel, mechanically strong, light weight composite bone implants. The results suggest that mechanical reinforcement is dependent on the nanostructure morphology, defects, dispersion of nanomaterials in the polymer matrix, and the cross-linking density of the polymer. In general, two-dimensional nanostructures can reinforce the polymer better than one-dimensional nanostructures, and inorganic nanomaterials are better reinforcing agents than carbon based nanomaterials.

Delivery from Compartmented Nanotubes

Nano tubes are also used for carrying drugs in general therapy and in tumor therapy in particular. The role of them is to protect the drugs from destruction in blood stream, to control the delivery with a well-defined release kinetics, and in ideal cases, to provide vector-targeting properties or release mechanism by external or internal stimuli.

Rod or tube-like, rather than nearly spherical, nanocarriers may offer additional advantages in terms of drug delivery systems. Such drug carrier particles possess additional choice of the axial ratio, the curvature, and the "all-sweeping" hydrodynamic-related rotation, and they can be modified chemically at the inner surface, the outer surface, and at the end planes in a very selective way. Nanotubes prepared with a responsive polymer attached to the tube opening allow the control of access to and release from the tube. Furthermore, nanotubes can also be prepared showing a gradient in its chemical composition along the length of the tube.

Compartmented drug release systems were prepared based on nanotubes or nanofibres. Nanotubes and nanofibres, for instance, which contained fluorescent albumin with dog-fluorescein isothiocyanate were prepared as a model drug, as well as super paramagnetic nanoparticles composed of iron oxide or nickel ferrite. The presence of the magnetic nanoparticles allowed, first of all, the guiding of the nanotubes to specific locations in the body by external magnetic fields. Super paramagnetic particles are known to display strong interactions with external magnetic fields leading to large saturation magnetizations. In addition, by using periodically varying magnetic fields, the nanoparticles were heated up to provide, thus, a trigger for drug release. The presence of the model drug was established by fluorescence spectroscopy and the same holds for the analysis of the model drug released from the nanotubes.

Immobilization of Proteins

Core shell fibers of nano particles with fluid cores and solid shells can be used to entrap biological objects such as proteins, viruses or bacteria in conditions which do not affect their functions. This effect can be used among others for biosensor applications. For example, Green Fluorescent Protein is immobilized in nanostructured fibres providing large surface areas and short distances for the analyte to approach the sensor protein.

With respect to using such fibers for sensor applications fluorescence of the core shell fibers was found to decay rapidly as the fibers were immersed into a solution containing urea: Urea permeates through the wall into the core where it causes denaturation of the GFP. This simple experiment reveals that core–shell fibers are promising objects for preparing biosensors based on biological objects.

Polymer nanostructured fibers, core–shell fibers, hollow fibers, and nanorods and nanotubes provide a platform for a broad range of applications both in material science as well as in life science. Biological objects of different complexity and synthetic objects carrying specific functions can be incorporated into such nanostructured polymer systems while keeping their specific functions vital. Biosensors, tissue engineering, drug delivery, or enzymatic catalysis is just a few of the possible examples. The incorporation of viruses and bacteria all the way up to microorganism should not really pose a problem and the applications coming from such biohybrid systems should be tremendous.

Engineering Applications

Polymer nanocomposites are important for the automotive tire industry due to the possibility of achieving a higher fuel efficiency by designing polymer nanocomposites with suitable properties.

The most common type of filler particles utilized by the tire industry had traditionally been Carbon black (Cb), produced from the incomplete combustion of coal tar and ethylene. The main reason is that the addition of Cb to rubbers enables the manufacturing of tires of a smaller rolling resistance which accounts for about 4% of the worldwide CO_2 emissions from fossil fuels. A decrease in the rolling resistance of the car tires produced worldwide is anticipated to decrease the overall fuel consumption of cars due to the fact that a vehicle with tires of a smaller rolling resistance requires less energy to thrust forward. However, a smaller rolling resistance also leads to a lower wet grip performance, which poses concerns of the passenger's safety.

The problem can be partially solved by replacing Cb with silica, because it enables the production of "green" tires that display both improved wet grip properties as well as a smaller rolling resistance.

The main difference in the relevant properties of Cb and silica is that Cb is hydrophobic (as are the polymers used in the manufacturing of car ties) whereas silica is hydrophilic. So in order to increase the compatibility among the silica fillers and the polymer matrix, the silica is usuallyfunctionalized with coupling agents, which gives the possibility of tuning the filler-polymer interactions and thus producing nanocomposites of specific properties.

Overall, the main unresolved issue on the mechanical properties of filled rubbers is the elucidation of the exact mechanism of their mechanical reinforcement and of the so-called Payne effect; and owing to a lack of suitable theoretical and experimental approaches, both of them are still poorly understood.

Size and Pressure Effects on Nanopolymers

The size- and pressure- dependent glass transition temperatures of free-standing films or supported films having weak interactions with substrates decreases with decreasing of pressure and size. However, the glass transition temperature of supported films having strong interaction with substrates increases of pressure and the decrease of size. Different models like two layer model, three layer model, T_g (D, 0) ∝ 1/D and some more models relating specific heat, density and thermal expansion are used to obtain the experimental results on nanopolymers and even some observations like freezing of films due to memory effects in the visco-elastic eigenmodels of the films, and finite effects of the small molecule glass are observed. To describe T_g (D, 0) function of polymers more generally, a simple and unified model recently is provided based on the size-dependent melting temperature of crystals and Lindemann's criterion,

$$T_g(D, 0) / T_g(\infty, 0) \propto \sigma_g^2(\infty, 0) / \sigma_g^2(D, 0)$$

where σ_g is the root of mean squared displacement of surface and interior molecules of glasses at T_g (D, 0), $\alpha = \sigma_s^2(D, 0) / \sigma_v^2(D, 0)$ with subscripts s and v denoting surface and volume, respectively. For a nanoparticle, D has a usual meaning of diameter, for a nanowire, D is taken as its diameter, and for a thin film, D denotes its thickness. D_0 denotes a critical diameter at which all molecules of a low-dimensional glass are located on its surface.

Polymer Layered Silicate Nanocomposites

Polymer nanocomposites, especially polymer-layered silicate nanocomposites, represent a radical alternative to conventionally (macroscopically)filled polymers. Because of their nanometer-size dispersion, the nanocomposites exhibit markedly improved properties when compared with the pure polymers or conventional composites. These include increased modulus and strength, decreased gas permeability, increased solvent and heat resistance and decreased flammability.

For example, a doubling of the tensile modulus and strength is achieved for nylon-layered silicate nanocomposites with as little as 2 vol% of inorganic content. In addition, the heat distortion temperature of the nanocomposites increases by up to 100 °C, extending their use to higher-temperature environments, such as automotive under-the-hood parts. Further-more, the heat release rate in the nanocomposites is reduced by up to 63% at heat fluxes of 50 kW m^{-2} without an increase in the CO and soot produced during combustion.

Synthesis of Nanocomposites

Melt intercalation of high polymers is a powerful new approach to the synthesis of polymer-layered silicate, (PLS) nanocomposites. This method is quite general and is broadly applicable to a range of commodity polymers, from essentially non-polar polystyrene to weakly polar poly(ethylene terephthalate) to strongly polar nylon. PLS nanocomposites can thus be processed using current technologies, and easily scaled to manufacturing quantities.

In general, two types of hybrid structures are possible: Intercalated, in which a single, extended polymer chain is intercalated between the silicate layers, resulting in a well-ordered multilayer with alternating polymer/inorganic layers; and disordered or delaminated, in which the silicate layers (1) nm thick) are exfoliated and dispersed in a continuous polymer matrix.

Structure and Properties of Layered Silicates

The silicates used belong to the general family of so-called 2:1 layered silicates. Their crystal structure consists of layers made up of two silica tetrahedra fused to an edge-shared octahedral sheet of either alumina or magnesia. Stacking of the layers leads to a regular van der Waals gap between the layers called the interlayer or gallery. Isomorphic substitution within the layers generates negative charges that are normally counterbalanced by cations residing in the interlayer.

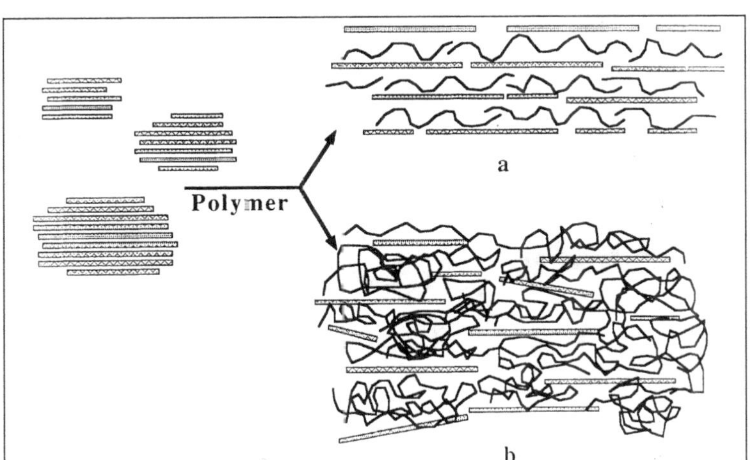

Schematic of composite structures obtained using layered silicates. The rectangular bars represent the silicate layers. (a) Single polymer layers intercalated in the silicate galleries (intercalated hybrids); (b) composites obtained by delamination of the silicate particles and dispersion in a continuous polymer matrix (delaminated hybrids).

Pristine layered silicates usually contain hydrated Na^+ or K^+ ions. Ion-exchange reactions with cationic surfactants, including primary, tertiary and quaternary ammonium ions, render the normally hydrophilic silicate surface organophilic, which makes intercalation of many engineering polymers possible. The role of the alkylammonium cations in the organosilicates is to lower the surface energy of the inorganic component and improve the wetting characteristics with the polymer. Additionally, the alkylammonium cations can provide functional groups that can react with the polymer or initiate a polymerization of monomers to improve the strength of the interface between the inorganic component and the polymer.

Properties of Nanocomposites

Polymer-layered silicate nanocomposites exhibit many advantages, including the following: (a) They are lighter in weight than conventionally filled polymers because high degrees of stiffness and strength are realized with far less high-density inorganic material: (b) they exhibit outstanding diffusional barrier properties without requiring a multipolymer-layered design, allowing for recycling; and (c) their mechanical properties are potentially superior to unidirectional fiber-reinforced polymers, because reinforcement from the inorganic layers will occur in two dimensions rather than in one.

Recent work has also shown that the nanocomposites have excellent flammability properties. The heat release rate measured by cone calorimetry is reduced by 63% in a nylon-6 silicate nanocomposite containing 2 vol% of silicate. In addition, in contrast to many commercial fire retardants, the rate of soot and CO generation does not increase during combustion. The flame-retardant behavior of the nanocomposites has been attributed to the formation of a tough char layer which may act as a mass-transport barrier, slowing the escape of volatile by-products during combustion.

Similar improved char yields and flame resistance have also been observed by us for polycaprolactone,polyether imide and vinyl esterlayered silicate nanocomposites .

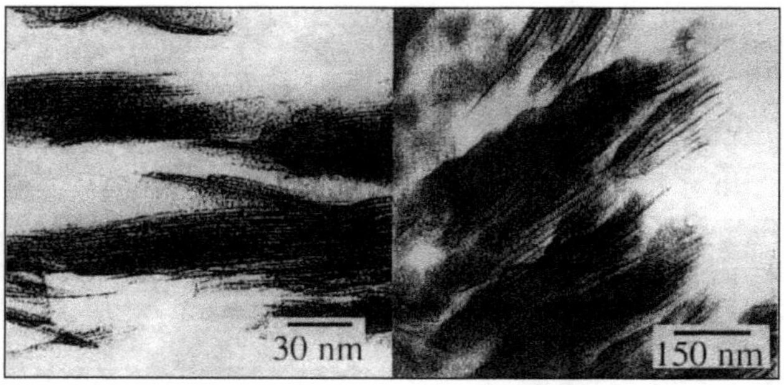

Figure Transmission electron micrographs of a polystyrene intercalated nanocomposite (left) and an epoxy delaminated nanocomposite (right).

For example, polyether imide nanocomposites retain about 90 and 45% of their weight after 20 and 120 min, respectively, at 450 °C. The respective values for the pure polymer are 45 and 15%. In the 500 °C isothermal the polymer is completely lost after 40 min, while the char yield of the nanocomposite is ca 55%. In addition, when samples are exposed to an open flame, burning of the pure polymer persists after the flame is removed, until it is externally extinguished. In contrast, the nanocomposites stop burning after the flame is removed and become highly charred but maintain their initial dimensions. It is important to note that the nanodispersion of the silicate is essential to this flame-retardant behavior. Polymers macroscopically filled with the same silicate as in the nanocomposites and at equivalent loadings show at best the same, and in some cases an increase (rather than a decrease), in polymer flammability.8 In contrast to other flame-retardant additives which tend to deteriorate in mechanical properties, the nanocomposites show the opposite trend. Some of the properties of nylon-6 nanocomposite containing only 2 vol% silicate (similar to the sample used in the cone calorimeter experiments). Typically, the same improvements in properties of conventionally filled nylon require additions of ca 30 vol% mineral.

In addition to the mechanical properties, the barrier properties of the nanocomposites are dramatically improved compared with the pure or macroscopically filled polymers, even at much higher filler content. The dramatic reduction in permeability has been attributed to the presence of well-dispersed silicate layers with a large aspect ratio which causes solutes to follow a tortuous path, thereby increasing the effective path length for diffusion.

Miscibility/Thermodynamics of Nanocomposite Formation

Similarly to polymer blends, any mixture of polymer and layered silicate does not necessarily lead to a nanocomposite. In most the incompatibility of the hydrophobic polymer and the hydrophilic silicate leads to phase separation similar to that of macroscopically filled systems. In contrast, by using surface-modified silicates, one can fine-tune their surface energy and render them miscible (or compatible) with different polymers. The approach is based on a chemical (rather than a mechanical) driving force which leads to nanoscopic dispersion.

What is then the driving force for polymer intercalation which leads to nanodispersion? To answer this question we have developed recently a mean-field model. Nanocomposites are formed if the free energy change, ΔG, of the process is negative. The change in free energy is composed of an enthalpic term, ΔH, due to the intermolecular interactions and an entropic term, ΔS, associated with the configurational changes of the constituents, and $\Delta G = \Delta H - T\Delta S$, where T is the temperature. The entropy change of the organically modified inorganic component is calculated using a modified Flory–Huggins lattice model in which the occupation of the lattice is weighted to simulate the preferred orientations of the organic modifier (tethered chains) in the presence of two impenetrable surfaces (silicate layers). The confinement of the polymer chains is similarly approximated using a self-consistent field treatment of a random-flight polymer between two surfaces. For the enthalpic term a modified mean-field, site-fraction approach where the number of contacts per lattice site is replaced by an interaction area per lattice site is being used. This approach makes it possible to express the interaction parameter as energy per area and may be approximated by interfacial or surface energies.

Table: Properties of nylon-6 and silicate–nylon nanocomposites.

Property	Nanocomposite	Nylon-6
Tensile modulus (GPa)	2.1	1.1
Tensile strength (MPa)	107	69
Heat distortion temp. (°C)	145	65
Impact strength (kJ m^{-2})	2.8	2.3
Water adsorption (%)	0.51	0.87
Coefficient of thermal expansion	6.3×10^{-5}	13×10^{-5}

Using this model we find that the entropy loss associated with the polymer confinement is approximately compensated by an entropy gain associated with the increased conformational freedom of the surfactant molecules as the gallery distance increases due to the polymer intercalation. Therefore, enthalpy determines whether or not polymer intercalation will take place. As the polymer–silicate interactions become stronger, first intercalated and then delaminated nanocomposites are predicted, in agreement with the experiment.

Structure and Dynamics of Polymer Nanocomposites

In addition to their potential applications, PLS nanocomposites are unique model systems to study the statics and dynamics of polymers in confined environments.10 Generally, the dynamics and structure of polymers in confined environments are a fundamental aspect of many industrially important fields such as tribology, adhesion, lubrication, sorption and catalysis. Using both delaminated and intercalated hybrids, the statics and dynamics of polymers confined over distances ranging from the radius of gyration of the polymer to the statistical segment length of the chains can be studied.

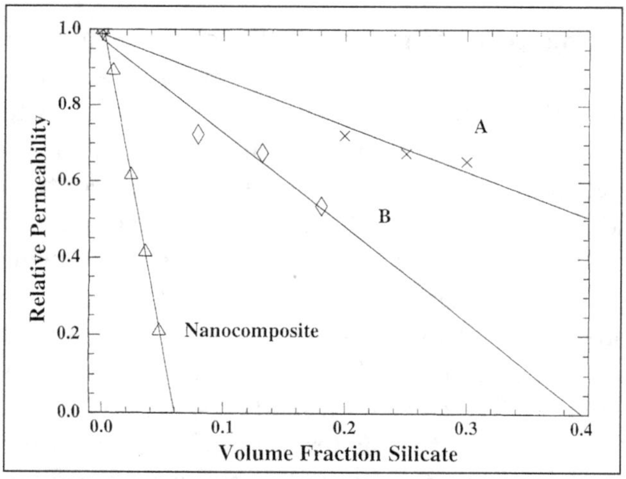

Relative permeability as a function of inorganic content of (A) polycaprolactone nanocomposites and (B) conventionally filled (i.e. without nanodispersion) polymers.

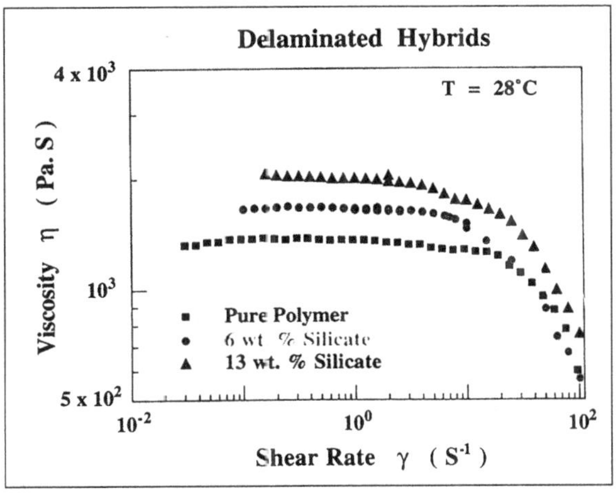

Steady-state shear viscosity as a function of shear rate for polydimethylsiloxane layered silicate nanocomposites containing different levels of silicate loadings.

In general, the behavior of polymer liquids under confinement is very different from that in the bulk, especially when the polymer thickness approaches the size of the polymer coil. Traditional notions such as reptation dynamics governing the relaxation of a long polymer chain are improbable in highly confined intercalated systems (confinement distances comparable with the statistical segment length of the polymer), as it is difficult to imagine an entanglement in two dimensions.

Despite the topological constraints imposed by the host lattice, mass transport of the polymer into the silicate layers appears to be unhindered and exhibits mobility which is faster than the selfdiffusion of the pure polymer. The observation that polymer chains can undergo center-of-mass transport in essentially two dimensions is rather surprising because the unperturbed radius of gyration of the polymer is roughly an order of magnitude greater than the interlayer distance between the silicate layers. The ability of the chains to undergo center-of-mass transport during melt intercalation is evidence that the interactions within the interlayer does not completely restrict segmental motions, otherwise large-scale chain motion would not be possible.

In addition, the dynamics of the polymer in the hybrids are dramatically different from those in the bulk. On a local scale, intercalated polymer chains exhibit higher flexibility along their backbone, together with a marked suppression (or even absence) of the cooperative dynamics typically associated with the glass transition.

In contrast to macroscopically filled polymers, in which filler addition leads to dramatic increases in viscosity, silicate nanocomposites exhibit shear thinning. For example, the steady-state shear viscosity of a series of polydimethylsiloxane nanocomposites shows an increase with respect to that of the pure polymer at low shear rates but still obeys Newtonian-type behavior, even at reasonably high silicate loadings figure. The same effect is observed in small amplitude oscillatory shear measurements, wherein the storage and loss moduli for the nanocomposites display similar frequency dependence to the pure polymer, with a monotonic increase in the magnitude of the moduli with increasing silicate content. Although at present the exact mechanism that causes shear thinning is not known, we suspect that it is due to the orientation of the silicate layers under shear and polymer coils. Nevertheless, because of this shear thinning the nanocomposites can be melt-processed, and this simplifies their manufacturing.

Carbon Nanotube-based Polymer Composites

Carbon nanotubes (CNTs) are tubes made of carbon with diameters typically measured in nanometers.

A scanning tunneling microscopy image of a single-walled carbon nanotube.

Carbon nanotubes often refers to single-wall carbon nanotubes (SWCNTs) with diameters in the range of a nanometer. They were discovered independently by Iijima and Ichihashi and Bethune et al. in carbon arc chambers similar to those used to produce fullerenes. Single-wall carbon nanotubes are one of the allotropes of carbon, intermediate between fullerene cages and flat graphene.

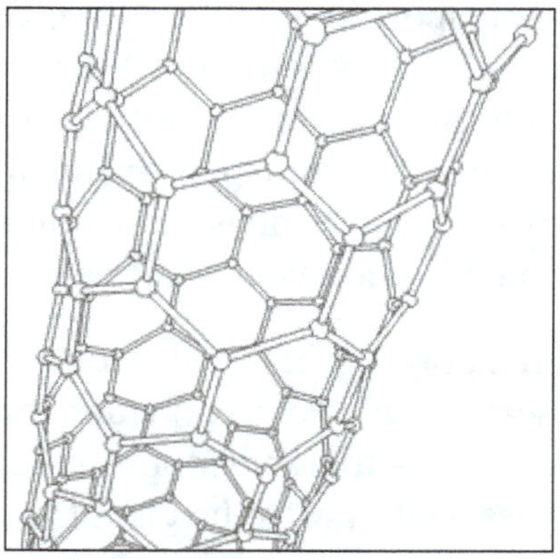

Rotating single-walled zigzag carbon nanotube.

Although not made this way, single-wall carbon nanotubes can be thought of as cutouts from a two-dimensional hexagonal lattice of carbon atoms rolled up along one of the Bravais lattice vectors of the hexagonal lattice to form a hollow cylinder. In this construction, periodic boundary conditions are imposed over the length of this roll up vector to yield a lattice with helical symmetry of seamlessly bonded carbon atoms on the cylinder surface.

Carbon nanotubes also often refer to multi-wall carbon nanotubes (MWCNTs) consisting of nested single-wall carbon nanotubes. If not identical, these tubes are very similar to Oberlin, Endo and Koyama's long straight and parallel carbon layers cylindrically rolled around a hollow tube. Multi-wall carbon nanotubes are also sometimes used to refer to double- and triple-wall carbon nanotubes.

Carbon nanotubes can also refer to tubes with an undetermined carbon-wall structure and diameters less than 100 nanometers. Such tubes were discovered by Radushkevich and Lukyanovich. While nanotubes of other compositions exist, most research has been focused on the carbon ones. Therefore, the "carbon" qualifier is often left implicit in the acronyms, and the names are abbreviated NT, SWNT, and MWNT.

Carbon nanotubes can exhibit remarkable electrical conductivity. They also have exceptional tensile strength and thermal conductivity, because of their nanostructure and strength of the bonds between carbon atoms. In addition, they can be chemically modified. These properties are expected to be valuable in many areas of technology, such as electronics, optics, composite materials (replacing or complementing carbon fibers), nanotechnology, and other applications of materials science.

Individual carbon nanotubes naturally align themselves into "ropes" held together by relatively weak van der Waals forces. The length of a carbon nanotube produced by common production methods is often not reported, but is much larger than its diameter. Although rare, nanotubes half a meter long have been created, with a length-to-diameter ratio of more than 100,000,000:1. For many purposes, the length of carbon nanotubes can be assumed to be infinite.

Rolling up a hexagonal lattice along different directions to form different single-wall carbon nanotubes shows that all of these tubes have helical and translational symmetry along the tube axis and many also have nontrivial rotational symmetry about this axis. In addition, most are chiral, meaning the tube and its mirror image cannot be superimposed. This construction also allows single-wall carbon nanotubes to be labeled by a pair of small integers.

A special group of achiral single-wall carbon nanotubes are metallic, but all the rest are either small or moderate band gap semiconductors. These electrical properties, however, do not depend on whether the tube is rolled up above or below the graphene plane and hence are the same for a tube and its mirror image.

Structure of Carbon Nanotubes

The electronic configuration of carbon atom is $1s^2\ 2s^2\ 2p^2$ indicating that it has two strongly bound electrons in the 1s orbital and four relatively weakly bound electrons in 2s and 2p orbitals known as valence electrons. The small energy difference between 2s and 2p levels allows the carbon atom to exist in several hybridization states; sp, sp^2 and sp^3 in different materials.

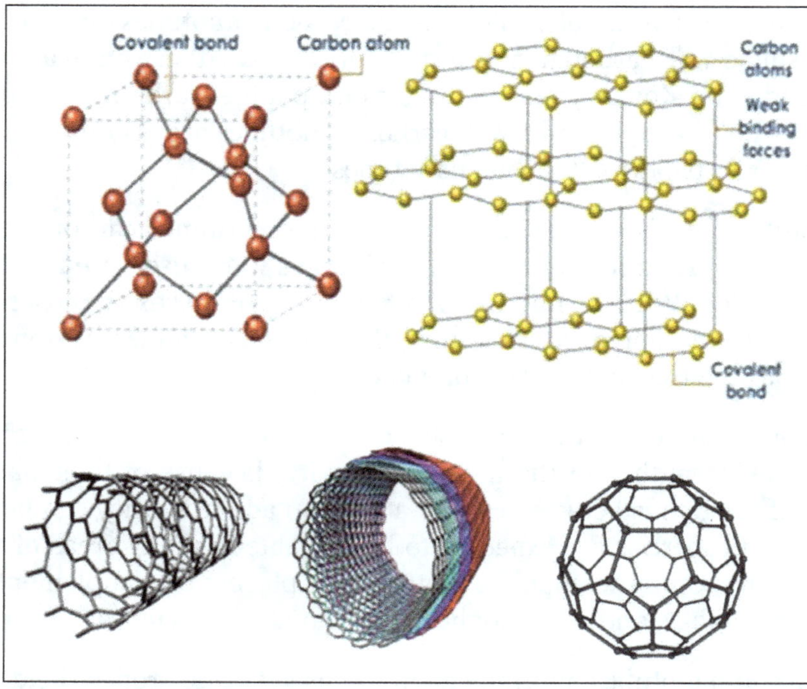

Schematic diagrams of diamond, graphite, fullerene single-wall carbon nanotube (SWCNT) and multi-wall carbon nanotube (MWCNT).

The hybridization flexibility enables the atomic orbitals of carbon to arrange themselves in structures of different dimensionalities ranging from diamond (3D), graphite (2D), carbon nanotubes (1D) and fullerene (0D). CNTs are hollow cylinders of graphene with extraordinary electronic and mechanical properties. They exist in two varieties viz. CNTs composed of a single graphene sheet called single-wall carbon nanotubes (SWCNTs) and an array of coaxial nanotubes known as multiwall carbon nanotubes (MWCNTs).

In the honeycomb lattice, the vector C_h is called chiral vector, while T is called the translational vector. The rectangle generated by the chiral vector C_h and the translational vector T is the unit cell of the SWCNT in real space. The chiral angle is defined as the angle between the chiral vector C_h and the unit vector a_1. The chiral vector can be expressed in terms of the unit vectors a_1 and a_2 by means of integers (n, m) (which are called chiral indices):

$$h=n_1+m_2$$

The wrapping of graphene sheet is governed by the different orientations of the chiral vector leading to the different CNT geometries. When the chiral indices are equal (n = m), the SWCNT is called armchair, and the chiral angle is 30°. If one of the chiral index is zero (n, 0) or (0, m), the SWCNT is named zig-zag, and in this case, the chiral angle is 0° (achiral nanotubes). In the other cases (n ≠ m), the nanotube is called chiral and its chiral angle is $0° < \theta < 30°$.

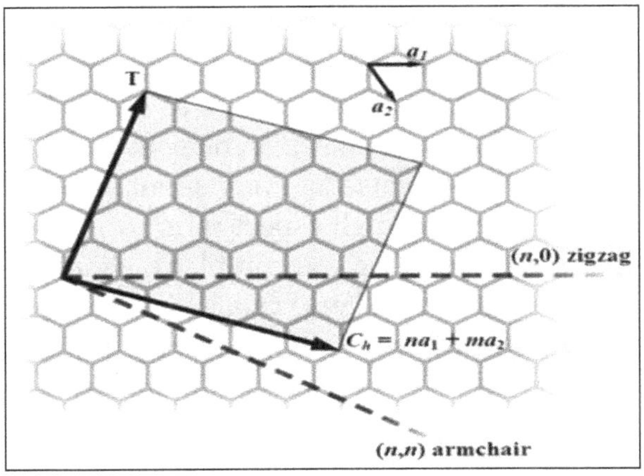

Chiral vector Ch and chiral angle θ definition for a nanotube on graphene sheet. a1 and a2 are the unit cell vectors of the two-dimensional hexagonal graphene sheet.

The expressions for the main parameters of a tube as a function of the chiral indices (n, m) are:

$$\text{nanotube diameter } D = a\sqrt{(n^2 + m^2 + nm)}$$

$$\text{with, } a = a_{C-C}\sqrt{3} = 1.44\,\text{Å}$$

$$\text{Chiral angle } \theta, \cos\theta = \frac{(2n+m)}{2\sqrt{(n^2 + m^2 + nm)}}$$

Length, Diameter and Aspect Ratio

The actual length of CNTs is expected to influence their physical status as well as acquired configuration inside polymer matrix. In general, long-length CNTs show a flexible fibre-like character with tube buckling and entangled configuration. Therefore, its processing inside polymer matrix is a bit difficult in terms of de-entanglements. The long length ensures lower electrical percolation threshold, good mechanical properties due to the large-distance stress transfer and crack propagation prevention ability. In contrast, the short-length CNTs display rod-like character, provide more stiffness (compared to long-length CNTs of similar loading) to composite, and has relatively large percolation threshold, inferior stress transfer, poor crack countering property, low tensile/flexural strength and elongation of composites. However, it can be easily aligned and tends to show better interfacial polarization. Like the length of CNTs, diameters also affect the properties, for example SWCNTs, large diameter offers large interfacial surface area per unit length but less stiffness. Similarly, for MWCNTs, large diameter means more number of cylindrical graphitic tubular shells (i.e. thick walls), which tends to improve inherent mechanical properties of tubes, its stiffness, electrical properties and ability to undergo functionalization without much harm in tube's properties. Another important parameter is aspect ratio (i.e. length/diameter ratio), which affects the properties via interplay of length and diameter. However, it could be bit elusive and care has to be taken while talking about aspect ratio because a short-length and small-diameter CNT can have same aspect ratio as a long-length and large-diameter CNT, but their properties and translated effect on composites may be altogether different. Nevertheless, CNTs are widely used as very small, high aspect ratio conductive additives for plastics of all types. The high aspect ratio (up 1000 or more) gives CNTs an edge over other conductive fillers (e.g. carbon black, chopped carbon fibre, carbon nanofibres, stainless steel fibres or whiskers) as a lower loading of CNTs is needed to achieve the same electrical conductivity. This low loading preserves polymer resin's toughness, especially at low temperatures and maintains other key properties of the matrix resin.

Defects

A perfect single-wall CNT is only a theoretical construction and even perfect hexagonal sp^2 structure possesses different types of defects. Generally, a defective site has a high reactivity meaning that at those points, the chemical reactions are favoured. A simple example of a defect is the presence of non-hexagonal-shaped carbon rings such as a pentagon and heptagon pair (also known as Stone/Wales defect or 5/7 defect). They are localized mainly at tube ends and near tube bending locations. A Stone/Wales defect has very significant role to play in the modification of density of states of nanotubes which has implications for possible nanodevice applications. Defects are very important because they can modify the electronic properties of the nanotubes thereby influencing their applications. The distribution of the 5/7 defects in the capping of the CNTs can induce the presence of new localized states in the valence and/or the conduction band, and so to modify the field emission properties. The introduction of a 5/7 defect in the hexagonal structure can induce a local deformation of the diameter of the tube, a change of the chirality or the formation of a CNT intramolecular junction. The defects can also be present along the sidewall, and after acid treatment, it is possible to open nanotube ends and attach chemical functional groups facilitating the interaction of CNTs with other moieties. In addition to being important for the potential applications of nanotubes, defects also play a key role in the functionalization process.

Synthesis Methods of Carbon Nanotubes

Arc Discharge Method

The carbon arc-discharge method was first brought to light by Krätschmer et al. who utilized it to achieve the production of fullerenes in macroscopic quantities. Iijima discovered the catalyst-free formation of multiwall carbon nanotubes (MWCNTs), while investigating the other carbon nanostructures formed along with the fullerenes and more particularly, the solid carbon deposit forming onto the cathode. Figure shows the schematic diagram of the arc-discharge method. This method creates CNTs through arc-induced vapourization of two carbon rods placed end to end, separated by approximately

1 mm, in an enclosure usually filled with inert gas (argon or helium) at low pressure. A direct current of 50–100 A, driven by a potential difference of approximately 20 V, creates a high-temperature discharge between the two electrodes, vapourizes the surface of one of the carbon electrodes, and forms a small rod-shaped deposit on the other electrode. This technique produces a complex mixture of components including non-tubular forms of the carbon such as nanoparticles, fullerene-like structures including C_{60}, multiwall shells, amorphous carbon, etc.. Therefore, this requires further purification to separate the CNTs from the soot and the residual catalytic metals present in the crude product. In this technique, the high yield production of CNTs depends on the uniformity of the plasma arc, and the temperature of the deposit forming on the carbon electrode.

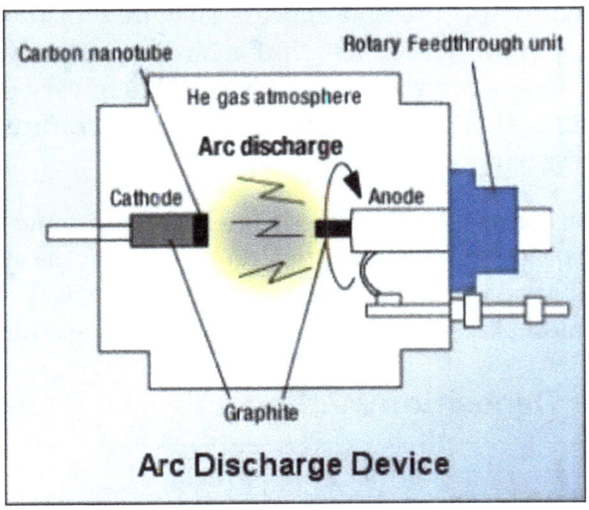

Schematic diagram of Arc-discharge setup.

Laser Ablation Method

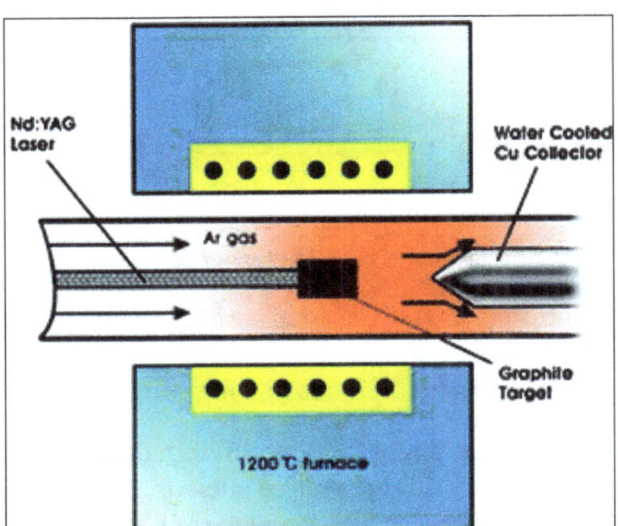

Schematic diagram of laser ablation setup.

The laser ablation technique was used successfully to synthesize fullerene for the first time in 1985 by Kroto et al.

Figure shows the schematic diagram of laser ablation method. Samples are prepared by laser vapourization of graphite rods with a 50:50 catalyst mixture of cobalt and nickel at 1200 °C in flowing argon, followed by heat treatment in a vacuum at 1000 °C to remove the C_{60} and other fullerenes. The target is vapourized more uniformly by a second laser pulse. The use of two successive laser pulses minimizes the amount of carbon deposited as soot. The second laser pulse breaks up the larger particles ablated by the first one and feeds them into the growing nanotube structure.

The material produced by this method appears as a 'mat' of ropes, 10–20 nm in diameter and up to 100 μm or more in length. Each rope primarily consists of a bundle of SWCNTs, aligned along a common axis. It is possible to vary the average nanotube diameter and size distribution by varying the growth temperature, the catalyst composition and other process parameters.

Both these solid carbon source-based synthesis methods have some drawbacks. One issue is the scaling up of the process to the industrial level. Secondly, the synthesized CNTs have impurities of metal catalyst particles and unwanted carbon forms such as fullerenes, amorphous carbon, multiwall shells, single-wall nanocapsule and need further purification.

Chemical Vapour Deposition Method

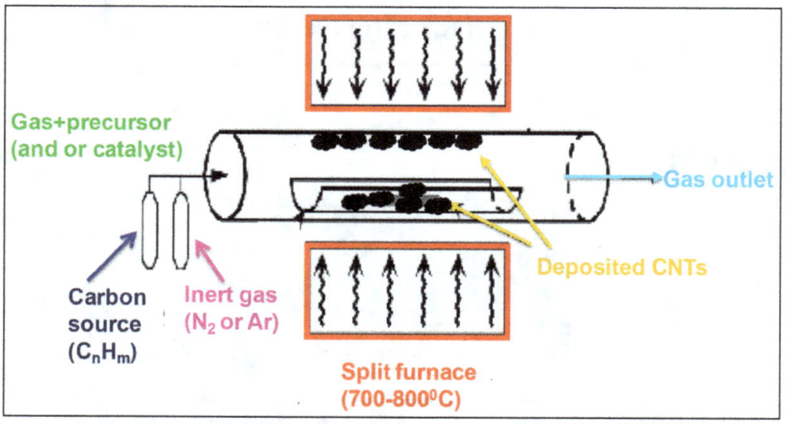

Schematic diagram of chemical vapour deposition setup.

Chemical vapour deposition (CVD) of gaseous carbon source (hydrocarbons, CO) over a metal catalyst is a classical method that has been used to produce various carbon materials such as carbon fibres and filaments for a long time.

Figure shows the schematic diagram of CVD method. In this technique, a carbon source is taken in the gas phase and an energy source such as plasma or a resistively heated coil is employed to transfer energy to a gaseous carbon molecule. The CVD technique uses hydrocarbons such as methane, carbon monoxide or acetylene as carbon source.

During CVD, a substrate covered with metal catalysts (such as nickel, cobalt, iron or their combination) is heated to approximately 700 °C. The growth starts after two gases are passed through the chamber, that is hydrocarbon gas and the other a carrier gas such as nitrogen, hydrogen or argon. The advantages of the CVD process are low power input, low-temperature range, relatively high purity and most importantly, the possibility to scale up the process. This technique can produce both MWCNTs and SWCNTs depending on the temperature, wherein production of SWCNTs occurs at a higher temperature than MWCNTs.

Functionalization of Carbon Nanotubes

Functionalization enhances the solubility and processibility, and allows combining the unique properties of nanotubes with those of other materials. It also improves the interaction of the nanotube with other entities, such as a solvent, polymer and other organic molecules and also with other nanotubes. A functionalized nanotube displays different mechanical and electrical properties as compared to pristine nanotube and thus may be utilized for several applications.

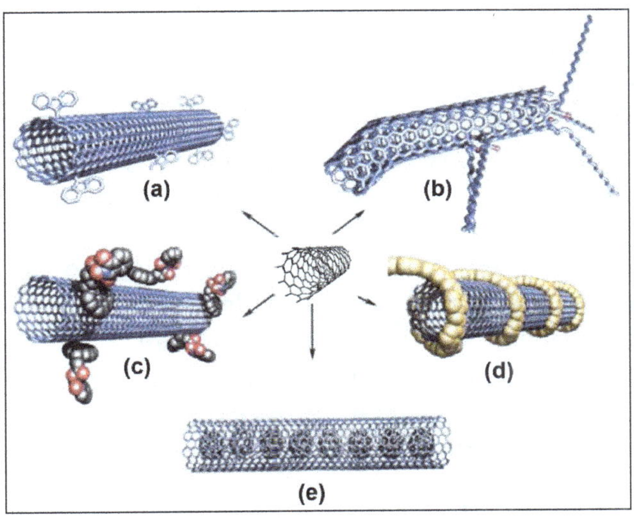

Different possibilities of the functionalization of SWCNTs: (a) Sidewall functionalization. (b) Defect-group functionalization. (c) Non-covalent exohedral functionalization with molecules through π-stacking. (d) Non-covalent exohedral functionalization with polymers. (e) Endohedral functionalization, in this case C60@SWCNT.

The process of functionalization can conveniently be divided into three major types depending upon the chemistry involved.

Covalent Functionalization

Covalent functionalization utilizes the covalent linkage of functional entities onto the nanotube's carbon scaffold. Depending upon the site of interaction, it can be of two types—covalent sidewall functionalization and defect functionalization. Direct covalent sidewall functionalization involves a change of hybridization from sp^2 to sp^3 and the loss of conjugation. Defect functionalization is based on the transformations of defect sites already present. Defect sites can be the open ends and holes in the sidewalls, terminated, for example by functional groups and Stone/Wales defects (5–7 defects) in hexagonal graphene framework. In addition to these, oxidative purification generated oxygenated sites are also considered as defects. SWCNTs demonstrate low dispersability, and they occur in the form of bundles. This situation warrants the use of a highly reactive reagent for the covalent bond formation at the sidewalls. It is not possible to tell beforehand that whether these addition reactions are more likely to take place at defect sites or intact hexagonal regions of sidewall. Several covalent routes have been taken for covalent functionalization such as oxidative purification, amidation, esterification, thiolation, halogenation, hydrogenation], cycloadditions, electrochemical functionalization.

Non-covalent Functionalization

Covalent functionalization suffers from the drawback of damaging the structure of CNTs. Therefore, with a view to retain the structural integrity and π network of CNTs, non-covalent functionalization is particularly attractive. The adsorption forces, such as van der Waals' and π-stacking interactions, are responsible for the non-covalent interaction between surface active reagents and CNTs. Broadly, surfactants and polymers are utilized for this non-destructive functionalization of CNTs. Surface active molecules such as sodium dodecylsulfate (SDS) or benzylalkonium chloride have been successfully used for the formation of non-covalent aggregates. On the other side, polymer wrapping around CNTs can be accomplished through the use of polymers such as poly (m-phenylene-co-2, 5-dioctoxy-p-phenylenevinylene) (PmPV, 2) in organic solvents such as chloroform. The stable solution of the SWNT/PmPV complex exhibits conductivity eight-times higher than that of PmPV, without any compromise on its optical properties. The polar side-chain polymers such as polyvinylpyrrolidone (PVP) or polystyrenesulfonate (PSS) give stable solutions of the SWNT/polymer complexes in water.

The covalent and non-covalent functionalizations are essentially exohedral derivatizations.

Endohedral Functionalization

The hollow inner cavity of SWCNTs serves as a capillary for the storage of nanoparticles and fullerenes, etc. The rich endohedral chemistry of SWCNTs is amply illustrated by the incorporation of fullerenes and metallofullerenes in their cavity.

CNTs-based Polymer Composites

The spectacular properties of CNTs such as their high strength and stiffness make them ideal candidates for structural applications. At present, polymer nanocomposite is one of the biggest application areas for CNTs. The extraordinary properties of CNTs coupled with easily tailorable characteristics of polymers give rise to truly versatile CNT-polymer nanocomposites. The emergence of CNTs as filler materials has contributed in the realization of CNT-polymer nanocomposites as next generation advanced structural material. Keeping in view, the ubiquitous need for the creation of materials with tailored properties for various applications, it is hardly an exaggeration to call the present age as the 'age of composites'.

A composite is a multiphase material formed from a combination of materials which differ in composition or form, retain their own chemical and physical properties, and maintain an interface between components which act in concert to provide improved specific or synergistic characteristics not obtainable by any of the original components acting alone. Any composite is composed of two categories of materials: the reinforcement or filler and the matrix. The reinforcements contribute useful properties (mechanical, electrical, thermal, optical, etc.) to enhance the matrix properties. In the past decades, we have reached the technological design limits of optimizing composites with traditional micro-meter scale fillers/reinforcements. The limitations of traditional micrometer-scale polymer composites have prompted a considerable research effort to focus on polymer nanocomposites. This new hopes-laden advancement over the traditional polymer composites is characterized by the presence of at least one phase which is less than 100 nm in at least one dimension. Historically, nanocomposites are not entirely new as some nanocomposites such as carbon black and fumed silica-filled polymers have been used a long time ago. However, the discovery of CNTs added an additional impetus in the polymer nanocomposite research promising hitherto unknown potential for various applications. Table shows a comparison between various fibre reinforcements and carbon nanotubes in terms of various parameters.

Table: Comparison and contrast of advanced fibre reinforcements vs. carbon nanotubes in terms of diameter, density, tensile strength and modulus.

Fibre	Diameter (μm)	Density (g/cm³)	Tensile strength (GPa)	Modulus (GPa)
Carbon	7	1.66	2.4–3.1	120–170
S-glass	7	2.50	3.4–4.6	90
Aramid	12	1.44	2.8	70–170
Boron	100–140	2.50	3.5	400
Quartz	9	2.2	3.4	70
SiC fibres	10–20	2.3	2.8	190
SiC whiskers	0.002	2.3	6.9	–
Carbon NTs	0.001–0.1	~1.33	Up to ~50	Up to ~1000

Nanocomposite Fabrication Methods

A variety of synthesis techniques are in practice for incorporation of CNTs into various polymeric host matrices. The main motive is to deagglomerate the CNTs and realize their uniform dispersion inside polymer matrix. Currently, there is no single technique which is universally applicable to all the situations. Depending upon the thermal or chemical properties of matrix polymer, ease of its synthesis from suitable monomer, desired performance indices of composites and cost constraints, with some trade-off in properties, one can choose the suitable processing method for a particular case.

Solution Processing

The solution processing is the most common technique to form CNT-based polymer nanocomposites which exploits intensive agitation (e.g. refluxing, mechanical/magnetic stirring, vigorous shaking, high shear homogenization, bath/probe sonication) aided rigorous and thorough mixing of CNTs with polymer in a solvent, so as to facilitate nanotube de-bundling and their dispersion inside host polymer matrix. It is important to note that this technique is limited to the polymers which are soluble in solvent(s).

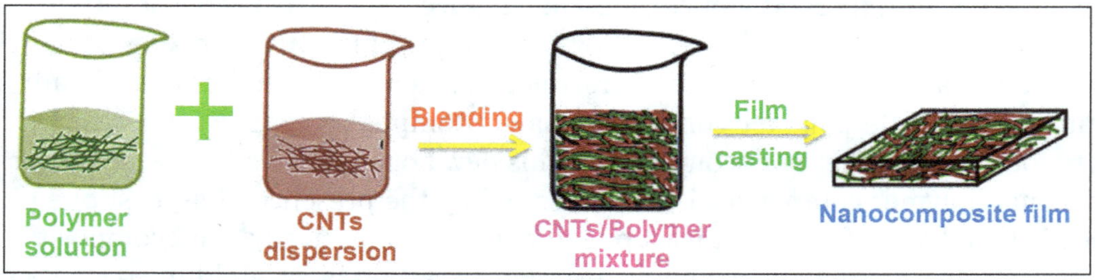

Schematic representation of solution processing method.

Typical process involves dispersion of nanotubes in a suitable solvent and mixing with the polymer solution, followed by film casting and solvent evaporation leaving behind nanocomposite film/sheet. The choice of solvent is mainly governed by solubility of matrix polymer. Further, the solvent for CNTs and polymer dispersion may be same or different, but should be of good missibility to realize intimate mixing between phases. In many cases, CNTs are not separately dispersed rather they are directly added to polymer solution followed by intensive mixing before film casting. The major shortcoming associated with the usage of high-power ultrasonication or shear mixing for a long time is that it can lead to the shortening of tube lengths, thereby deteriorating the composite properties. Plenty of literature is available on the formation of CNT-based nanocomposites by this method using both organic as well as aqueous media and variety of polymer matrices. To enable better dispersion and to solve the problem of tube shortening upon high-power agitation, some efforts have been made to use surfactants for tube dispersion or to use physically/chemically functionalized CNTs. However, the use of surfactant carries the unavoidable limitation of retaining surfactant in the nanocomposites which hinders the thermal/electrical transport properties of nanocomposites. In contrast, the functionalized CNTs often provide positive results in terms of CNTs deagglomeration, dispersion and their improved interfacial adhesion with matrix polymer, which gets reflected in terms of superior electrical, thermal, mechanical and dielectric properties of resultant nanocomposites. The boiling point of dispersion medium or solvent (s) is found to have a profound impact on the properties of formed nanocomposites. In general, low-boiling point (i.e. fast evaporating or drying) solvents are preferred, due to their ease of removal from the solution-casted nanocomposites mass. In contrast, high-boiling point solvents are difficult to be removed and tend to get trapped in the solidifying/curing mass. Such a trapped solvent may interfere with the curing reaction (in thermosets) or can act as softener (in thermoplastics), thereby adversely affecting electrical, thermal or mechanical properties. In addition, efforts to remove the solvent and to prevent evaporation generated voids formation add towards complexity and cost in terms of requirement of systems for controlled heating, vacuum/pressure. Another, limitation encountered quite often with solution processing is that the slow evaporation of solvent provides sufficient time for CNTs re-agglomeration and differential settling, resulting in inhomogenous CNTs dispersion in matrix (e.g. CNTs content lowest at the casted film/sheet's surface, shows a uniform/random gradient across thickness and maximum at both surfaces due to the extensive tube settling) and observation of non-uniform and inferior properties. The solvent evaporation rate-related limitations can be resolved by gently pouring CNT/polymer nanocomposite dispersion on a rotating substrate (spin coating) or over a heated substrate (drop-casting). However, use of spin coating is limited only to thin films (few nanometres thick) which cannot be peeled off from the substrate, whereas drop casting has issues in terms of uniform drying across thickness and high possibility of void formation. Another versatile method exploits coagulation of CNT/polymer dispersion by pouring into an excess of non-solvent, thereby achieving rapid precipitation of polymer chains which immediately entrap CNTs (without providing sufficient time for CNTs diffusion and settling). Nevertheless, solution processing is still widely used and is one of the important steps in the processing of thermosetting matrices-based nanocomposites.

Melt Processing

The melt processing is considered as viable option for making thermoplastic matrices-based CNT/polymer composites due to its low cost and amenability towards large scale synthesis for industrial applications. Here, elevated temperatures are exploited to melt the receptor thermoplastic matrix polymers, which form a viscous liquid and made to flow. The molten polymer flow induces high shear forces which assist in partial de-agglomeration of CNTs bundles and their dispersion inside matrix. The melt mixing can be carried out in batch or continuous operation using high shear mixer (e.g. Sigma mixer) and extruder, respectively. Sigma mixer is often used to prepare highly concentrated nanocomposites called masterbatches, which maybe used to synthesize desired CNTs-loading composites by their mixing with proportionate amount of neat matrix polymer using extruder. An extruder may consist of single or twin screws with twin screw version being more effective in terms of mixing uniformity and properties. A typical twin screw extruder consists of two co- or counter-rotating screws inside barrel housing. The polymer granules are caught by the rotating screws and pushed forward; they become melted inside heated melting (feed) zone due to the externally provided heat and shearing of the material between screw and barrel. The CNTs are loaded into the extruder via separate hopper, such that melt-phase mixing takes place due to the combination of shearing and kneading action, and by the time molten-mixture reaches the homogenization zone, it has already achieved significant degree of mixing. Finally, mixture passes to die before coming out as semi-solid strands, which may be cooled (via air drying or by passing through water bath) and chopped into granules for further use, for example for compression moulding. The extruded output may also be diverted to injection moulding machine to form desired shape samples.

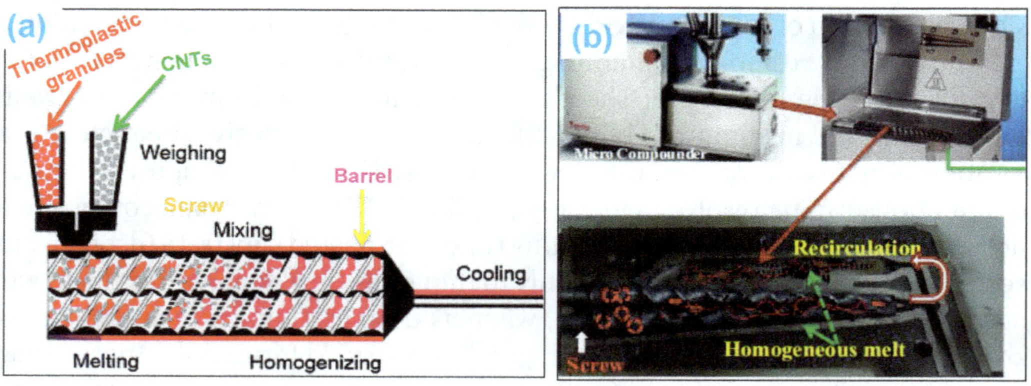

Schematic representation of twin screw extruder for melt phase mixing of CNTs with thermoplastic matrices. Reprinted from with permission from Wiley.
(b) Twin screw extruder with melt recirculation provision.

In the past, melt mixing/blending is successfully exploited for dispersion of CNTs inside various thermoplastic matrices, for example polystyrene, polypropylene and acrylonitrile-butadiene-styrene (ABS), polyamide-6, polyethylene. Although melt-mixing technique is simple but the issues of high shear force and elevated temperature need to be properly addressed in order to avoid the deterioration of nanocomposites. While high shear force facilitates CNT dispersion, it can also lead to the undesirable CNT fragmentation or even polymer chain scission. Therefore, the shear force needs to be optimized to achieve desired dispersion without compromising the structural integrity of CNTs. Similar situation prevails with high temperature which promotes CNTs dispersion but can degrade the intrinsic properties of polymer making the need of optimizing the temperature indispensable. Further, it is found that melt-mixing technique is not much effective in breaking of agglomeration of CNTs compared to solution processing. Besides, the melt-mixing fails to realize dispersion of high loading (>5 wt%, which are required for thermal/electrical/electromagnetic applications) inside polymeric matrices, due to the viscosity buildup and screw rpm/reside time limitations. Recently, the above issues are resolved by using a twin screw extruded equipped with melt recirculation provision, such that upto 10 wt% CNTs can be loaded inside polypropylene copolymer (PPCP) matrix without any deleterious effect on mechanical properties.

In Situ Polymerization

In situ polymerization remains the only viable option for the preparation of composites based on insoluble or thermally unstable matrix polymers, which cannot be processed by solution or melt processing routes. However, many a times, it is used in other cases too (where above limitations are not applicable), due to the distinguished advantages of in situ polymerization in terms of ability to allow formation of high CNTs-loading nanocomposites, facilitating good CNTs dispersion within polymer matrix and ensuring excellent intimacy between CNTs and matrix polymers.

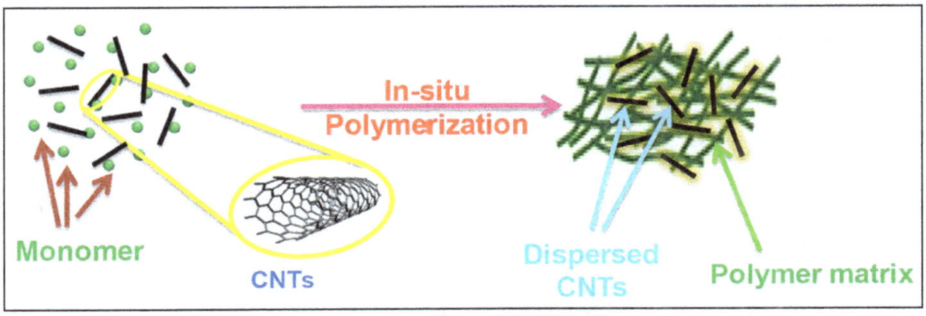

Schematic representation of in situ polymerization process.

This strategy involves dispersion of CNTs in monomerfollowed by in situ polymerization leading to the formation of CNT/polymer nanocomposites. Exploitation of functionalized CNTs or use of monomer-grafted CNTs are known to improve the initial dispersion of the nanotubes in the monomer and consequently in the formed nanocomposites. This results in a stronger and more active interface between nanotube and polymer which is pivotal to the nanocomposite performance for structural, electronic, electromagnetic or electrochemical applications. This method has been used to synthesize CNTs-filled composites with various polymers, for example thermoplastics, thermosets or conjugated polymers-based matrices. In addition to composite formation, in situ polymerization is also used for physical functionalization of CNTs (via surface polymer wrapping), for their onward use as hybrid filler for nanocomposites.

Miscellaneous Routes

In addition to aforementioned methods, some other less popular methods are also available for CNT-polymer nanocomposites preparation, for example wet spinning, batch mixing inside banbury mixer, twin screw pulverization, electrophoretic deposition and latex processing, twin screw extrusion complemented by melt-recirculation provision. The issue of the synthesis of CNT-based polymer nanocomposites is still an open arena as no single technique is entirely satisfactory on all grounds. Therefore, some efforts have also been made to use combination of techniques, for example solution processing with melt mixing; in situ polymerization with solvent processing; or in situ polymerization with melt processing.

Properties of CNT/Polymer Nanocomposites

With the advent of CNTs on technological landscape, there are now myriad possibilities for tailoring the properties of CNT/polymer nanocomposites. The electrical, thermal, dielectric, rheological and mechanical properties of CNTs filled nanocomposites are significantly enhanced compared to neat polymers. However, the issue of property enhancement is a challenging one as the improvement in one property might come at the cost of other. A number of factors such as the nature of matrix polymer, aspect ratio or actual length/diameter of CNTs, its pretreatment (e.g. covalent functionalization, surface coating of polymer or surfactant), loading level, exploited processing technique and the presence of tertiary phase (e.g. compatibilizer) are known to exert decisive influence on the properties of formed nanocomposites.

Electrical Properties

The very high intrinsic conductivity and low aspect ratio (length/diameter ratio) of CNTs compared to other carbon-based fillers (e.g. carbon black, graphite, carbon fibre), inspired researchers to synthesize CNTs filled electrically conducting nanocomposites. The incorporation of CNTs leads to onset of electrical conductivity within otherwise insulating matrix. This can be attributed to formation of 3D electrically conductive networks within host thermoset matrix so that electrons can easily hop/tunnel between dispersed filler particles. The minimum filler loading where first continuous network of filler particles were formed within matrix polymer is known as percolation threshold. At this point, electrical conductivity of composites displays a sharp rise, that is several orders of magnitude.

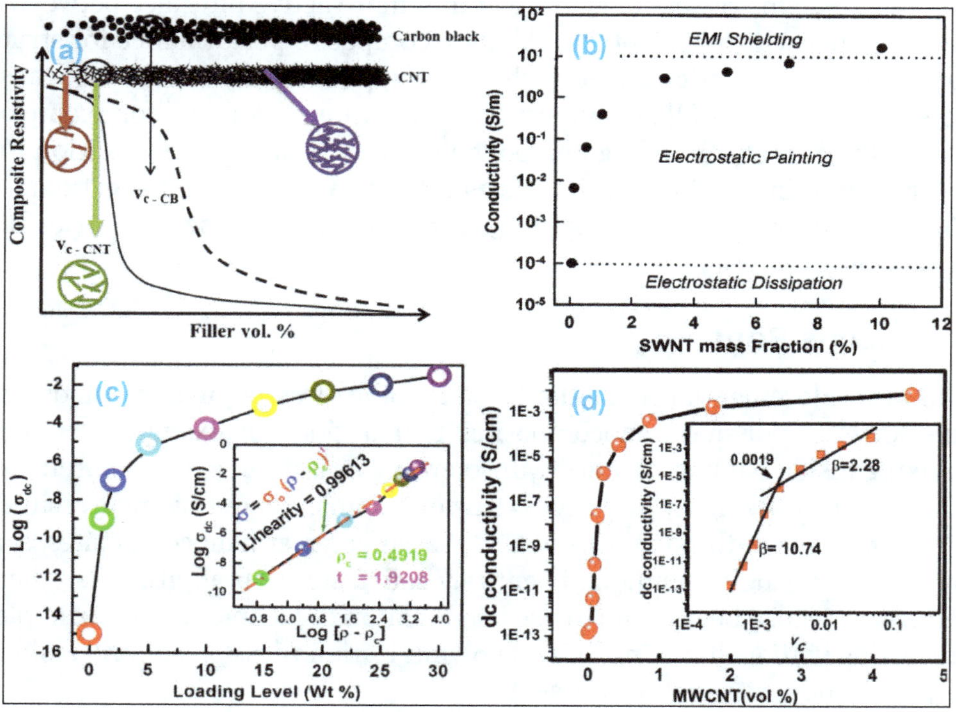

Previous figure shows, (a) Schematic representation of difference between percolation behaviour of CNTs compared to carbon black. (b) Electrical conductivity of SWNT/polycarbonate nanocomposites as a function of nanotube loading, showing a typical percolation behaviour. Dashed lines represent the lower limits of electrical conductivity required for the specified applications. (c) Variation of conductivity (σdc) of PANI-MWCNT nanofiller-loaded polystyrene solution blends. Inset shows the percolation and scaling details. (d) Plot of electrical conductivity (σ) versus MWCNT (vol%) for PPCP/MWCNTs composites. Inset shows the log–log plot of r as a function of (vc).

Depending upon the level of achieved electrical conductivity, these conductive nanocomposites may have a multitude of applications including electromagnetic interference (EMI) shielding, transparent conductive coating, electrostatic painting and electrostatic dissipation. Especially, electrically conductive CNT-polymer composites are used in anti-static packaging applications, as well as in specialized components in the electronics, automotive and aerospace sectors.

The critical conducting filler loading (i.e. percolation threshold) for a given matrix-filler combination can be calculated by plotting the electrical conductivity as a function of the reduced volume fraction of filler and performing data fitting with a power law function:

$$\sigma = \sigma_0 (v - v_c)^t$$

where σ is the electrical conductivity of the composite, σ_0 is characteristic conductivity, v is the volume fraction of filler, v_c is volume fraction at the percolation threshold, and t is the critical exponent. The log (σ) versus log ($v - v_c$) plot (inset) gives a straight line. In practical situations, where the densities of polymer matrix and filled inclusion are same (e.g. for organic fillers like ICPs, CNTs or graphene), the mass fraction, p, and volume fraction (v) of the filler can be assumed same. The percolation threshold can also be determined by plotting the log (σ) versus log (v) plot and finding the point of intersection of lines corresponding to different (β) values.

The CNT/polymer composites show very low percolation threshold for electrical conductivity. For SWCNT/polymer composites, the percolation thresholds ranging from 0.005 vol% to several vol% have been reported. On the other hand, in case of MWCNT/polymer composites percolation threshold up to 0.002 vol% has been reported. It is found that compared to common conductive fillers, for example metallic or graphitic particles in any shape (spherical, platelet-like or fibrous) and size, CNTs display much lower percolation threshold. This can be ascribed to combination of their high inherent conductivity and very high aspect ratio.

It is found that the percolation threshold in CNT/polymer nanocomposites depends on several parameters viz. dispersion, alignment, aspect ratio. Higher aspect ratio is obtained for well-dispersed nanotubes relative to nanotube bundles making percolation threshold decrease with better dispersion. Bryning et al. reported a smaller percolation threshold with the higher aspect ratio nanotubes for SWCNT/epoxy composites. The percolation threshold is also significantly affected by alignment of the nanotubes in the polymer matrix. In the alignment condition, there are fewer contacts between the tubes which results in a reduced electrical conductivity and a higher percolation threshold as compared to random orientation of nanotubes. It is a common notion that chemical functionalization reduces the electrical conductivity of nanotubes through disruption of the extended π-conjugation of nanotubes. However, it is also reported that functionalization can improve the electrical properties of the composites. Valentini et al. concluded that the amine-functionalized SWCNT in epoxy matrix allows migration of intrinsic charges raising the conductivity of the composite. Similar result is also obtained by Tamburri et al. for the functionalization of SWCNT with hydroxyl and carboxylic groups in 1,8-diaminophthalene. Therefore, it is established that negative effects of functionalization for SWCNTs conductivity are often counterbalanced by the improved dispersion caused by functionalization, with overall effect leading to positive outcome.

Thermal Properties

Due to the excellent thermal conductivity of CNTs, some efforts have been made to incorporate them into various polymer matrices to improve thermal conductivity of formed composites. It is found that, besides CNTs content, the thermal conductivity of CNT/polymer nanocomposites also depends on state of their dispersion and alignment as well as aspect ratio and the presence of metal impurities. Biercuk et al. synthesized an epoxy composite with 1 wt% raw laser-oven SWCNTs that showed a 125% increase in thermal conductivity at room temperature. Choi et al. reported a 300% increase in thermal conductivity at room temperature with 3 wt% SWCNTs in epoxy. They also obtained an additional 10% increment in case of magnetic alignment. Du et al. reported an infiltration method with an epoxy and a nanotube-rich phase showing a 220% increase in thermal conductivity at 2.3 wt% SWCNTs loading. Among various CNTs variants (SWCNTs, MWCNTs, DWCNTs) in epoxy composites, the MWCNTs are found to most significantly improve the thermal conductivity of polymer composites. This is due to their relatively low interfacial area (therefore, less phonon scattering at the interface) and the existence of shielded internal layers which promote the conduction of phonons and minimizes the matrix coupling losses. Besides, thermal conductivity improvement, CNTs are also known to improve the thermal stability of the polymer composites. This may be attributed to better heat dissipation characteristics. Nevertheless, these good thermal conductivity nanocomposites are considered very promising candidates for a number of applications such as thermal interface materials, heat sinks, printed circuit boards, connectors and other high-performance thermal management systems.

Mechanical Properties

The outstanding intrinsic mechanical properties of CNTs (including ultra-high-specific strength and modulus) make them especially lucrative as fillers in CNT-polymer nanocomposite materials. Due to these properties, only they were talked about as filler material for composite cables for NASA's fascinating project of space elevator. Therefore, several efforts have been made in the past, to translate the fraction of exceptional mechanical properties of CNTs into formed nanocomposites.

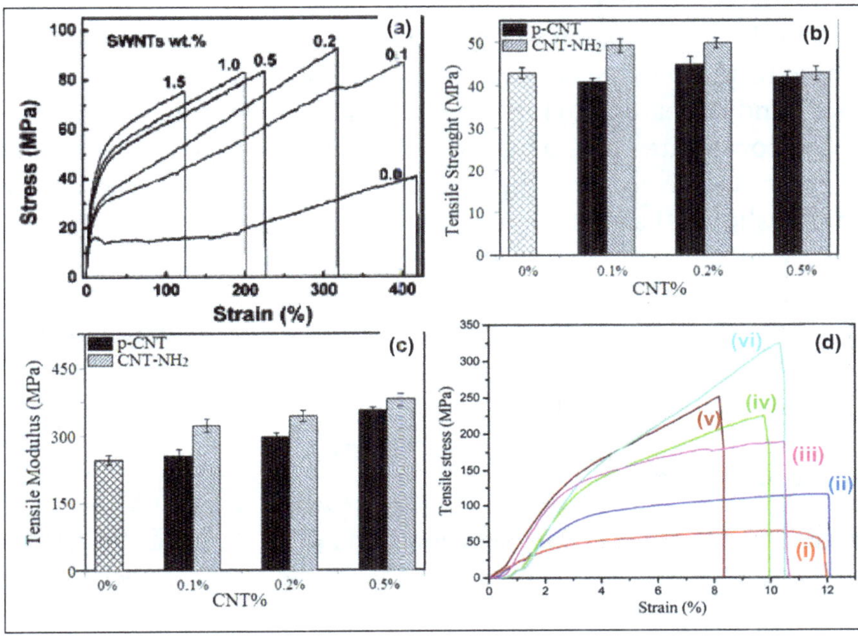

(a) Stress–strain profiles of SWNT-nylon-6 composite fibres at different SWCNT loadings. (b) Tensile strength and (c) tensile modulus of epoxy nanocomposites containing various % loadings of pure CNTs (p-CNT) and amino functionalized ($CNT\text{-}NH_2$). (d) Typical stress–strain curves for (i) electrospun pure PAN terpolymer nanofibre sheets; (ii) electrospun PAN terpolymer nanofibre sheets with 2 wt% loading of grafted MWCNTs; (iii) hot-stretched pure PAN terpolymer nanofibre sheets; (iv–vi) hot-stretched PAN terpolymer nanofibre sheets with 1, 2 and 3 wt% loadings of grafted MWCNTs, respectively.

In general, the tensile modulus and strength of CNT-polymer nanocomposites increase with nanotube loading, dispersion, and alignment in the polymer matrix However, unlike modulus, the tensile strength does not follow a monotonic increase with CNTs loading, due to CNTs re-aggregation, viscosity buildup issues, incomplete CNTs wetting by polymer and their poor dispersion. Therefore, optimum CNTs loading and means to overcome above negative factors are keys towards realizing composites with good mechanical properties. It is observed that the theoretical predictions of mechanical properties and experimental findings often differ due to the number of issues including the presence of voids, lack of perfect orientation, poor filler dispersion and insufficient load transfer due to the lack of interfacial adhesion. Moreover, nanotube agglomeration decreases the modulus of the nanotubes compared to that of isolated nanotubes because the nanotubes involve only weak dispersive forces between them. According to the molecular simulation studies and elasticity calculations done by Liao et al., when the atomic bonding between the nanotubes and the matrix is not present, there are basically two sources of nanotube/matrix adhesion (i) electrostatic and van der Waals interactions and (ii) stress/deformation resulting from the difference in the coefficients of thermal expansion between the filler and matrix. The functional groups on the nanotube sidewalls are known to improve the compatibility with the polymer matrix. This facilitates interfacial load transfer via CNT/polymer bonding leading to improved mechanical properties. The stress–strain profile of these composites give a 153 and 103% increase in Young's modulus and tensile strength, respectively. The importance of CNTs alignment has also been highlighted, and it is shown that the aligned CNTs-based nanocomposites display better strength and modulus in orientation direction compared to those based on random CNTs alignment. In spite of so many pros and cons, these lightweight yet structurally strong CNTs-based nanocomposites are considered as promising material for structural applications, for example bullet proof garments or body armours, aircraft or automobile parts, industrial components.

Dielectric Properties

It has been demonstrated that the incorporation of CNTs within insulating or conducting polymer matrices leads to improvement of dielectric properties. This can be attributed to the localization of charges at the CNTs/polymer interfaces, resulting in Maxwell–Wagner interfacial polarization. As a result, both real and imaginary permittivity of the polymer composites scales with CNTs loading.

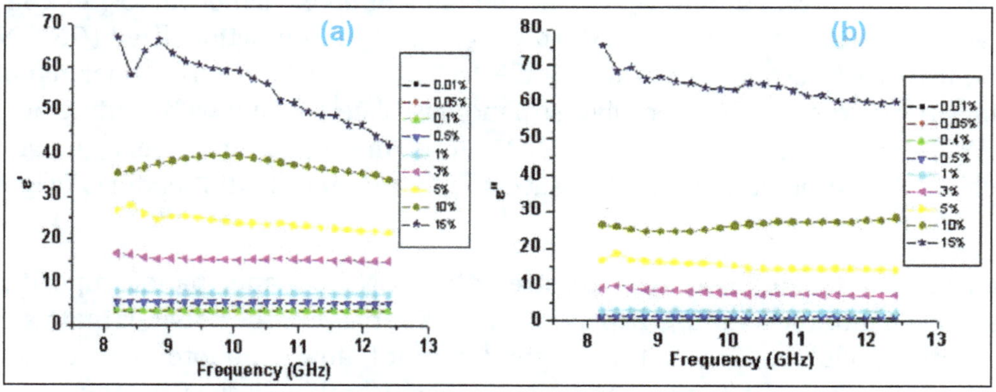

Complex permittivity spectra of the composites using "long-SWCNTs" with loading from 0.01 to 15 wt%.

The parameter ε ' (real permittivity) represents the charge storage (or dielectric constant), whereas ε " (imaginary permittivity) is a measure of dielectric dissipation or losses. It is found that dielectric properties are dependent on the CNTs' aspect ratio, actual length, functionalization/doping status and loading level as well as nature of polymeric matrix and employed processing technique. Such properties are best interpreted in terms of effective medium theory and have direct influence on electromagnetic wave-blocking characteristics of the CNTs/polymer nanocomposites.

Rheological Properties

The rheological properties of CNT/polymer nanocomposites have significance for composite processing as well as a probe of the composite dynamics and microstructure. At high frequencies, the response is almost independent of the filler concentration, indicating that the short-range polymer dynamics are not influenced by the nanotubes. The glass transition temperatures of the composites remain constant in the absence of strong interfacial bonds and for low nanotube loadings. At low frequencies, with the increase of nanotube concentration, the rheological behaviour gradually shifts from a liquid-like behaviour to a solid-like behaviour. This is in accordance with earlier findings in silicate nanocomposites. Applying a power law function to the data provides a rheological percolation threshold associated with the onset of solid-like response. As observed with respect to electrical percolation, the rheological percolation also depends on aspect ratio, nanotube dispersion and alignment. In a study by Mitchell et al., the effect of dispersion was demonstrated by functionalizing SWCNT such that the rheological percolation threshold decreased from 3 wt% when using pristine SWCNT to 1.5 wt% in functionalized SWCNT/polystyrene composites.

Pötschke et al. also showed the temperature dependence of the rheological percolation threshold. In SWCNT/ PC composite, the percolation threshold decreases from ~5 to ~0.5 wt% MWCNT when the temperature rises from 170 to 280 °C. Besides, elucidating structural information, rheological properties also provide useful information about the processability of the nanocomposites.

Application of CNT/Polymer Nanocomposites

Incorporation of CNTs inside various polymer matrices enables the formation of advanced nanocomposites with improved or even novel set of electrical, thermal, mechanical, electrochemical or electromagnetic properties.

Electromagnetic Interference (EMI) Shielding

The excellent electrical conductivity and high aspect ratio of CNTs compared to other carbon-based fillers (e.g. carbon black, graphite, carbon fibre) and their excellent corrosion resistance, low density along with ultra-high-specific strength compared to metals, inspired the designing of CNTs-based polymer composites. Both SWCNTs- and MWCNTs-based composites have been prepared by their incorporation in various polymer matrices (e.g. insulating thermoplastic or thermosetting polymers, conjugated polymers), and their EMI SE performance was measured. The introduction of CNTs inside host polymeric matrices leads to improvement of electrical conductivity as well as real- and imaginary-permittivity values. These are direct manifestations of increase in number of conducting links and interfacial polarization phenomenon and lead to improvement in SE.

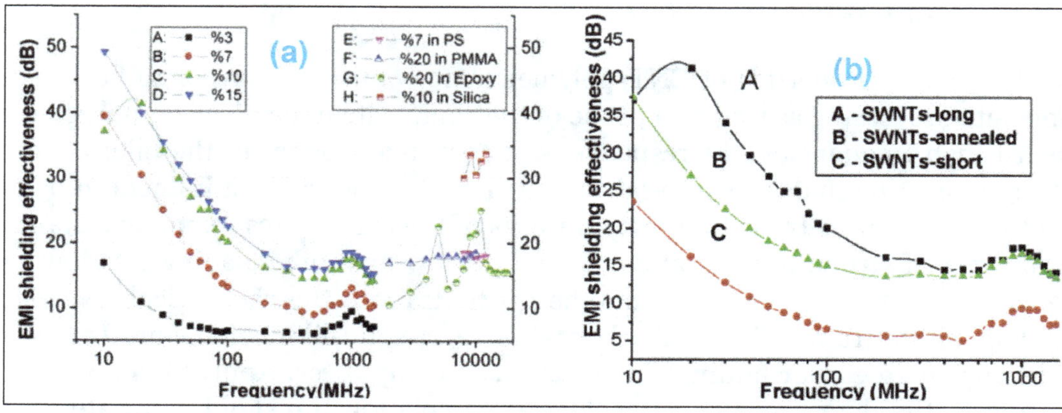

EMI shielding effectiveness (plots labelled A–D) for SWNT-polymer materials (wt% 3–15) studied in this work (10 MHz–1.5 GHz). Plots labelled (E–H) are higher-frequency data on MWNT-based material presented for comparison: (E) MWCNTs in PS; (F) MWCNTs in PMMA; (G) MWCNTs in epoxy resin and the value of the y axis for G is the reflection loss; (H) MWNTs in silica. Impact of wall integrity and aspect ratio on the EMI shielding effectiveness of the composites containing 10 wt% SWCNTs. Reprinted from with permission from ACS.

The SE value also depends upon the nature of polymer matrix, CNT-loading level and state of CNTs de-agglomeration/dispersion. Besides, CNT aspect ratio and wall defects also play crucial role in deterring shielding performance. In general, composites based on higher aspect ratio CNTs tend to display higher conductivities and better real and imaginary permittivity values compared to low aspect ratio CNTs, which gets reflected in terms of better shielding performance of the former. Further, annealed CNTs (containing very low defects)-based composites display superior shielding performance compared to unannealed CNT-filled composites.

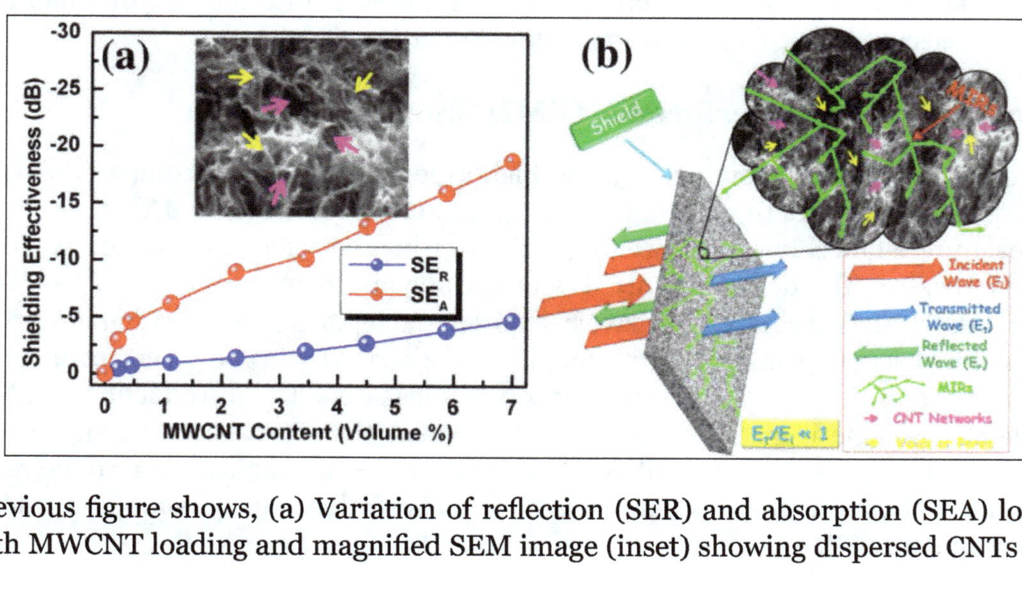

Previous figure shows, (a) Variation of reflection (SER) and absorption (SEA) losses with MWCNT loading and magnified SEM image (inset) showing dispersed CNTs and voids, (b) Schematic representation of radiation shield interaction and involved multiple internal reflection (MIR) phenomenon.

It has also been demonstrated that increased interfacial polarization and improvement of CNT dispersion via surface coating of conducting polymers lead to improvement of shielding response in the composites. It is suggested that CNTs networks inside composites trigger multiple reflections, which are considered beneficial for improvement of absorption and overall shielding. As the high CNT loading is required (where mechanical properties often degrade due to agglomeration and inhomogeneous filler dispersion issues) for achieving high shielding performance, some efforts have also been made to prepare high CNT loading (>10 wt%) yet mechanically strong composites via melt recirculation aided extrusive mixing. These composites display good shielding effectiveness. Recently, it is shown that at relatively low CNTs loading (<4 wt%), high aspect ratio and long-length CNTs-based composites display better shielding performance, whereas at higher (>4 wt%) CNTs loading, low aspect ratio and low-length CNTs-based composite show superior SE, probably due to the complex interplay between, impedance matching, interfacial polarization and multiple reflection phenomenon.

Supercapacitor Electrodes

CNTs, due to their superior electrical properties, good mechanical and thermal stability, readily accessible surface area and unique pore structure are an attractive candidate for supercapacitor electrode applications. The composite formation strategy, involving incorporation of CNTs into various conjugated polymers (e.g. PANI, PPY) matrix, is considered an effective solution for improving the mechanical and electrochemical properties of electrodes. For example, the sulfonated multiwall carbon nanotube (MWCNTs) in the composites can greatly improve the cycle stability of PANI (figure, left image), only showing 5.4% loss from their initial specific capacitance even after 1000 cycles. This was attributed to the use of MWCNTs with exceptional mechanical properties as a support and the formation of the charge-transfer complex, which could reduce the cycle degradation problems of PANI caused by volume changes or mechanical problems. Similarly, sulfonated-MWCNT/polypyrrole nanocomposite (MWCNTs/C-SO_3H/PPy) exhibited good rate ability, high-specific capacitance (357 F/g), and high-specific capacitance retention rate (specific capacitance loss was only 3% even after the 1000 cycles). Apart from preserving the conducting polymer active material from mechanical changes (such as shrinkage and breaking) during long cycling, CNTs also improved the charge transfer characteristics, thereby facilitating the realization of high charge/discharge rates.

It has also been realized that due to their excellent electrical and mechanical properties and open tubular mesoporous network structure, CNTs can as act good support materials for pseuocapacitive materials like conjugated polymers. Zhang and co-workers have reported the use of a carbon nanotube array directly connected to the current collector (Ta foil) as the support to make composite electrodes with hierarchical porous structures. The electrochemical studies have shown that these PANI/CNT composite electrode with nanosize hierarchical porous structure, large surface area, and superior conductivity had high-specific capacitance (1030 F/g), superior rate capability (95% capacity retention at current density of 118 A/g), and high stability (i.e. only 5.5% capacity loss after 5000 cycles) between potential window of −0.2 to +0.7 V (vs. SCE) in 1.0 M H_2SO_4 electrolyte. These results are considered as manifestations of efficient charge transport (in the composite electrode); high-specific capacity (due to the efficient utilization of electrode materials); improved ionic conductivity; and countering of mechanical stability problems or volume changes by CNTs. Some efforts have also been made to deposit conjugated polymer over CNTs-based membranes to be directly used in supercapacitor with metal backing. For example, pulsed electrochemically deposited PPy over MWCNT membrane-based electrode displayed remarkable specific capacitance of 427 F/g. Similarly, attempts have also been made to combine composite and paper electrode strategy, for example Oh et al. prepared the highly porous sheets comprised of SWCNT/PPy composite by vacuum filtration of SWCNT/PPy methanol dispersions. The nanotube network provided mechanical strength and improved electrochemical performance of the nanocomposites. The highest specific capacitance of 131 F/g was obtained for nanocomposite with 1:1 SWCNT: PPy ratio.

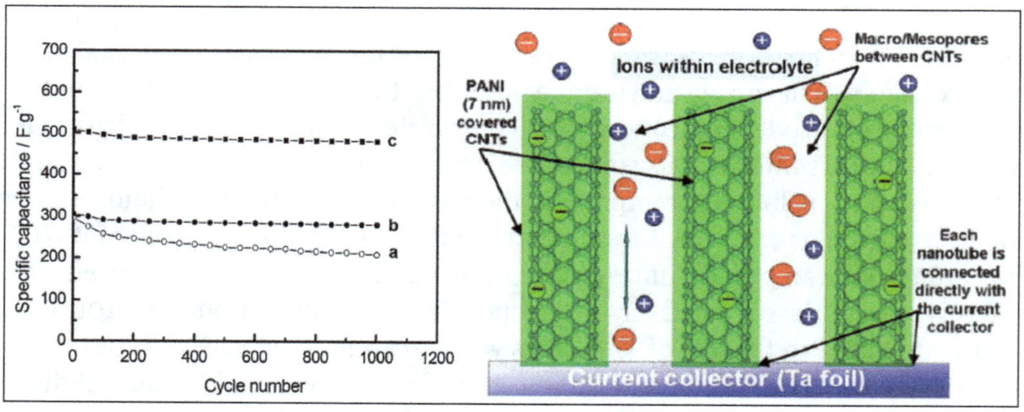

Cycle stability of (a) PANI nanorods, (b) PANI/sMWCNT-2, and (c) PANI/sMWCNT-4 electrodes in the voltage range of −0.2 to 0.75 V at a current density of 1 A/g (left image). Schematic representation of the microstructure and energy storage characteristics of a polyaniline/carbon nanotube (PANI/CNT) composite electrode (right image).

Photovoltaics

The excellent electron transport properties of carbon nanotubes (CNTs) and their larger electron affinity compared to available polymer-based light-harvesting donors prompted to use them in bulk heterojunction (BHJ)-type hybrid solar cells. Their acceptor characteristics have been demonstrated by their ability to rapidly quench (at very low CNTs loading) the photoluminescence (PL) of polymer-based donor which is a direct evidence of charge transfer from polymer (donor) to CNTs (acceptor). Further, CNTs also act as good transporter for exciton dissociation generated electrons. The initial effort in the direction demonstrated that incorporation of 1% SWCNTs inside poly(3-octylthiophene) (P3OT) donor lead to efficient exciton dissociation and improved open-circuit voltage (V_{OC}). Another study has highlighted the influence of SWCNTs on the crystallinity enhancement and morphology improvement (continuous interpenetrating network formation) in poly(3-hexylthiophene) (P3HT)-based solar cells. It is also outlined that workfunction of P3HT-modified SWCNTs increases due to the shifting of Fermi level towards vacuum level, leading to the improvement of V_{OC}. Another work showed that the solar cells BHJ active layer based on semiconducting SWCNTS (s-SWCNTs)/P3HT blend form nanofilaments which helps in preventing bundling of s-SWCNT via formation of core shell structures.

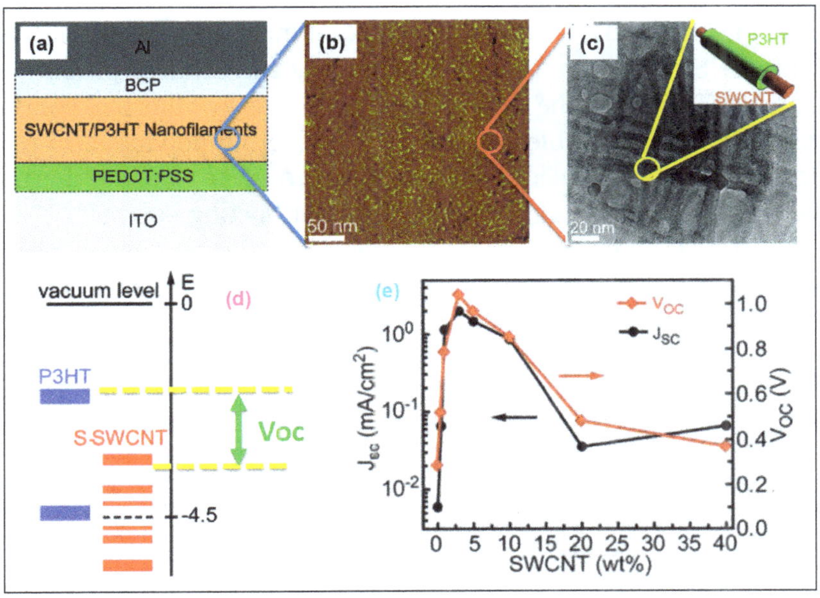

(a) Device structure of SWCNT/P3HT nanofilament BHJ layer-based solar cells. (b) AFM phase image of P3HT/s-SWCNT nanofilaments in a sample with 3 wt% SWCNT showing the worm-like morphology of the active layer. (c) Bright field TEM image of P3HT/s-SWCNTs blend having 3% s-SWCNT and schematic (inset) of P3HT coating over SWCNTs. (d) Band diagram for P3HT/s-SWCNTs interfaces and Voc. (e) Dependence of Jsc and Voc on the loading concentration of SWCNTs.

Such ordered configuration containing synergistic combination of s-SWCNTs and P3HT lead to improvement of charge separation, V_{oc} improvement and efficient carrier transport, which collectively contribute towards light-to-electricity conversion efficiency enhancement. The role of CNT content has also been investigated, and it was pointed that short-circuit current (Jsc) and V_{oc} of devices critically rely on the SWCNT loading. For example, in the present system, both the Jsc and V_{oc} display proportionate increases up to 3 wt% (V_{oc} of 1.04 V and Jsc of 1.99 mA/cm²) CNT content and decreases afterwards. Duly supported by morphological and electrical studies that was ascribed to the formation of optimal co-continuous interpenetrating-type morphology only around 3 wt% s-SWCNT loading. Above this value, phase segregation takes place, thereby decreasing the exciton breaking and charge transport efficiencies. It has also been shown that P3HT/s-SWCNT BHJs can generate photocurrent from photons absorbed both in the P3HT and in the s-SWCNT and can achieve an IQE of 26% in the near-infrared region. Another work demonstrated that incorporation of MWCNTs into the P3HT/C60 BHJ layer increased the fill factor by 20% with a corresponding improvement of efficiency compared to the polythiophene/C_{60} bilayer device containing no MWCNTs.

Thermoelectrics

Recently, several attempts have been made to combine the excellent electrical conductivity, tunability of Seebeck coefficient, outstanding mechanical properties and good thermal stability of CNTs with the solution processabilty, low thermal conductivity, cost advantages and facile and scalable synthesis of conducting polymers. In particular, nanoscale heterostructuring is considered as good approach to regulate thermopower and electrical/thermal conductivity. In early attempts, conducting polymer/CNTs nanocomposites (e.g. PANI/CNTs and PEDOT/CNTs) were synthesized by in situ polymerization to introduce heterojunctions and to see the effects of constituents on thermoelectric properties.

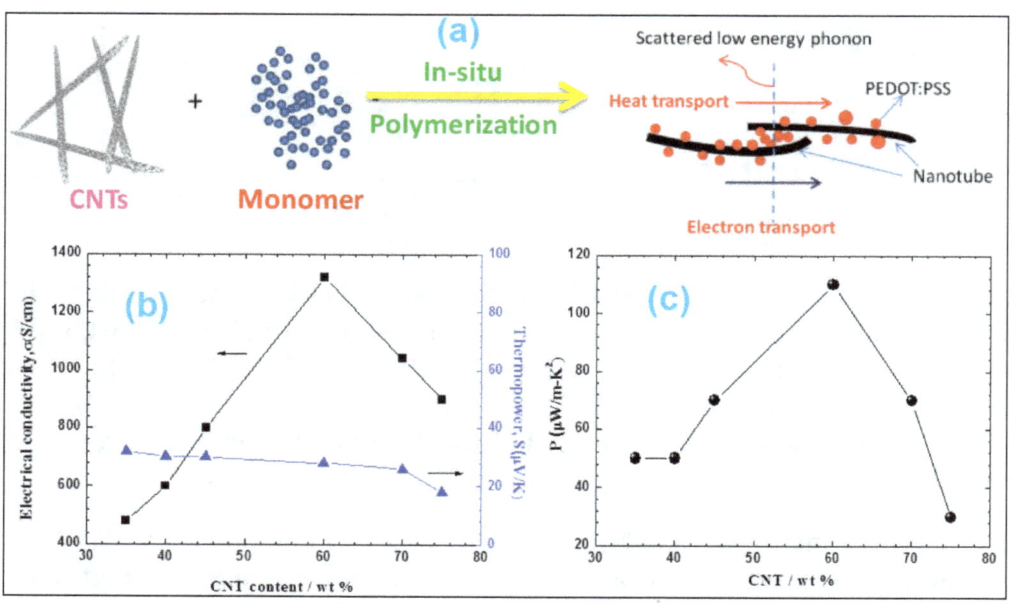

(a) Schematic representation of the formation of CNT/conjugated polymer by in situ polymerization leading to coating of nanotubes by polymer and formation of heat/electricity flow regulating nanojunctions. (b) Electrical conductivities and thermopower (i.e. seebeck coefficient) and (c) Power factor for the CNT/PEDOT-PSS composites.

The presence of the heterojunctions is expected to improve the thermoelectric transport properties, that is obstructing the heat flow and favouring the electronic conduction. Indeed, it has been observed that poly(3,4-ethylenedioxy-thiophene)-poly(styrenesulfonate) (PEDOT:PSS)/CNTs composites display improved electrical conductivity without significantly altering or decreasing thermopower (i.e. seebeck coefficient), ultimately resulting in improved power factor (PF). This behaviour results from thermally disconnected, but electrically connected junctions in the nanotube network, which makes it feasible to tune the properties in favour of a higher thermoelectric figure of merit. Similarly, when bundle of CNTs were coated and bounded by PANI, the thermally insulating PANI interfacial layer can act as energy filters, which allow the high-energy carriers to pass and scatter the low-energy carriers, thereby increasing the Seebeck coefficient. Furthermore, the growth of PANI over CNTs forms an ordered chain structure which reduced the π–π conjugation defects along the PANI backbone causing increase in carrier mobility (i.e. increased electrical conductivity). As a result, the composite possesses similar thermal conductivity as that of pure PANI (though orders of magnitude lower than pure CNTs) but significantly improved PF compared to pure PANI or CNTs.

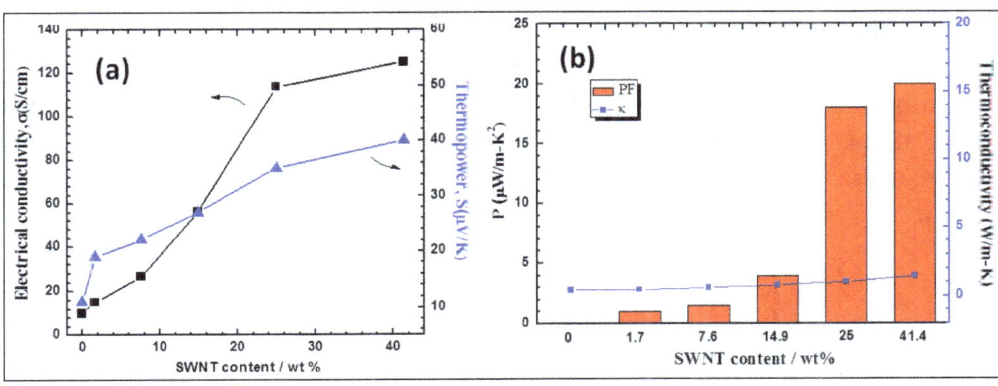

Seebeck coefficient and electrical conductivity (a), power factor and thermal conductivity (b) of SWCNT/PANI composites with different SWCNT content. The dashed line is the calculated electrical conductivity and thermal conductivity based on the particle mixture rule. Reprinted from with permission from ACS.

Besides ordering-induced improvement in carrier mobility, such systems also show anisotropic thermoelectric properties with more than a doubled improvement of power factor in the orientation direction. This provides a novel and effective way of improving the thermoelectric properties of conducting polymers.

In addition to conducting polymers, efforts have also been made to combine CNTs with conventional insulating polymers to improve thermoelectric PF. For example, in polyvinyli-dene fluoride (PVDF)/SWCNT composite thin films-based systems, due to the decrease in CNTs content (from 100 to 5 wt%), the beneficial effects of increasing Seebeck coefficient and decreasing thermal conductivity were outweighed by negative effect of decreasing electrical conductivity, resulting in an increase in a thermoelectric figure of merit (ZT). Similarly, CNT/poly(vinyl acetate) (PVAc) composites show highest thermoelectric performance at 20 wt% CNTs loading, with an electrical conductivity of 48 S/cm, thermal conductivity of 0.34 W/mK and room temperature ZT value larger than 6×10^{-3}. Besides, in addition to improvement in thermoelectric properties, due to their exceptional mechanical properties CNTs are also expected to improve the mechanical properties of their composite-based thermoelectric materials (both bulk materials and thin films).

Water Purification

In recent past, membrane-based filtration techniques have emerged as potential alternatives for waste water purification applications. Owing to their cost-effective facile synthesis, good thermal stability, high mechanical strength and biocompatibility, polymers such as polysulfone, polyamides, cellulose nitrate, polyethersulfone, membranes are the most promising/preferred. However, due to very small pore size, bacteriological contamination and pore blockings by adsorption of inorganic/organic impurities; low throughput and fouling are the common limitations of these membranes. In this context, CNTs owing to their strong antimicrobial activity, tunable surface chemistry and high mechanical strength have emerged as promising filler candidate for making composite membranes with improved antifouling characteristics and mechanical strength. Shawky et. al. has synthesized a MWCNT/polyamide nanocomposite membrane which exhibited excellent mechanical strength and very good salt rejection ability with high permeability.

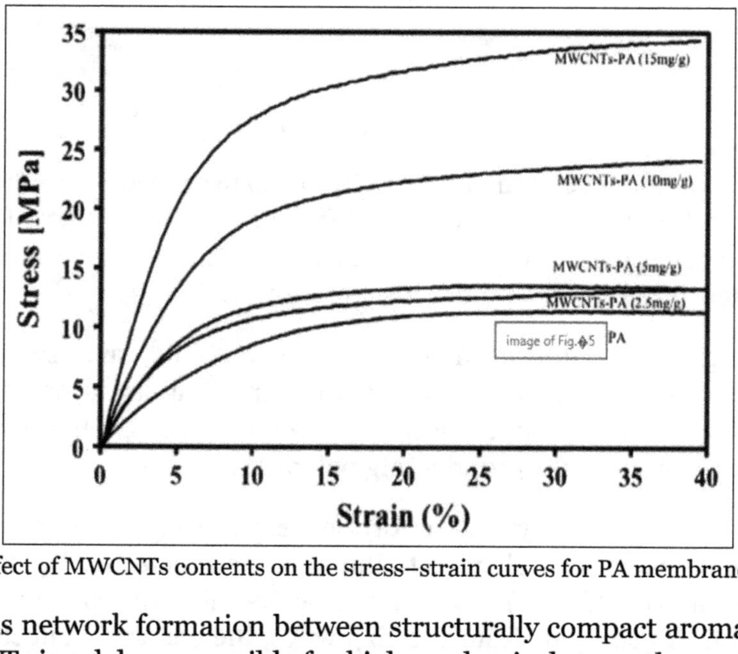

Effect of MWCNTs contents on the stress–strain curves for PA membranes.

The continuous network formation between structurally compact aromatic polyamide matrix and CNTs is solely responsible for high mechanical strength, good salt rejection ability though at the expense of slightly lower permeability.

Table: Membrane performance as a function of different MWCNTs loading at constant PA concentration (10%).

MWCNTs loading (mg/g)	Permeability ($L/m^2/h$ bar)	Flux (L/m^2 h)	Salt rejection (%)
0	0.76 ± 0.08	32 ± 0.7	24 ± 1.1
2.5	0.75 ± 0.09	32 ± 0.4	28 ± 1.0
5	0.73 ± 0.07	31 ± 1.1	35 ± 0.7
10	0.72 ± 0.10	30 ± 0.9	69 ± 0.9
15	0.71 ± 0.11	28 ± 0.8	76 ± 1.1

However, these membranes suffer from fouling tendency as the incorporation of hydrophobic CNTs in polyamide membranes leads to facile irreversible adsorption of organic/inorganic/biological impurities. It is shown that the fouling problems can be circumvented by altering the surface morphology of the membrane. Since CNTs can be easily transformed from hydrophobic to hydrophilic by acid treatment; therefore, incorporation of CNTs functionalized with hydrophilic/amphiphilic groups is considered very advantageous for long-term uninterrupted performance of theses membranes. In this direction, Choi et. al. have fabricated a polysulfone/MWCNT hybrid membrane by phase inversion process. The surface-modified MWCNTs provide hydrophilicity and conductivity to the membrane. The authors reported that the pore size of the membrane increased with increase in CNT loading up to 1.5 wt% but further loading of CNTs increased viscosity of blend solution which led to decrease in pore size. The membrane with 4 wt% of MWCNT has pore size just smaller than pure polysulfone membranes and exhibited higher flux and good salt rejection ability. Nevertheless, the CNTs reinforced polymeric membranes are still a new concept and more efforts in the direction are necessary to further improve their performance and address the negative influence on permeability.

Gas and Chemical Vapour Sensors

The major drawbacks such as high operating temperature, elevated costs and complex fabrication protocols associated with the conventional metal oxide-based sensors have prompted researchers to look for new avenues to come up with new materials not plagued by these shortcomings. This insatiable quest has unfolded the arena of CNTs-based polymer nanocomposites. The synergistic combination of the remarkable electrical transport and mechanical properties of CNTs and easily tailorable electroactive nature of conducting polymers has the potential to give one of the best-sensing platforms for the efficient gas/chemical vapour detection. Therefore, dedicated efforts have been made to combine CNTs with various conducting polymers such as polyaniline, poly(3,4-ethylenedioxythiophene)/poly(styrenesulfonate) (PEDOT/PSS), polypyrrole (PPy), hexafluoroisopropanol substituted polythiophene (HFIP-PT), poly(3-hexylthiophene) (P3HT), poly 3-methyl thiophene, to fabricate portable, more stable, highly sensitive, cost-effective and energy efficient chemiresistors especially for the detection of extremely minute quantities of environmentally hazardous analytes. In general, incorporation of CNTs tends to improve the mechanical strength, charge transport properties, porosity and specific interactions (due to high-specific surface area available for physico-chemical adsorption) with analytes (gas or chemical vapour), whereas conducting polymer contributes towards electroactivity and improved CNTs dispersion (thus better film formability compared to pristine CNTs). Therefore, the CNTs/conducting polymer nanocomposites often display improved sensitivity, response or selectivity.

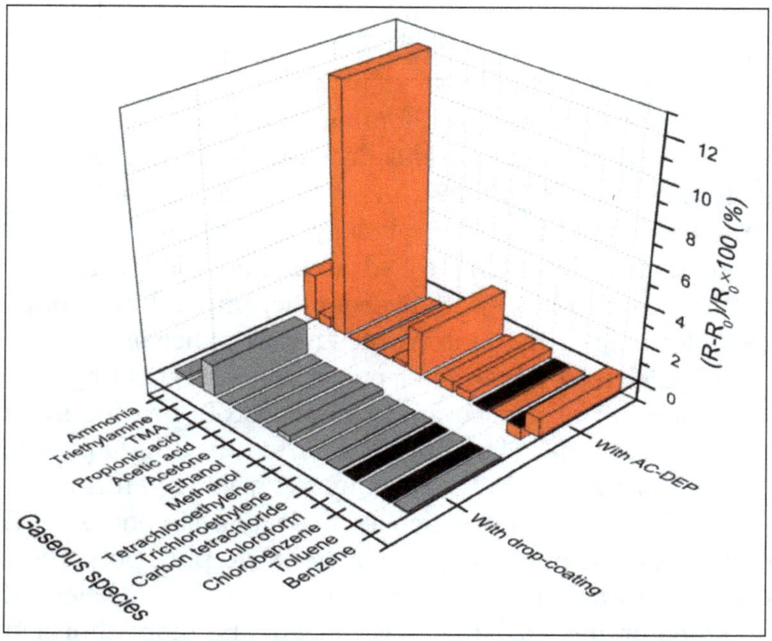

Selective response of the drop coated and AC-DEP assembled PEDOT/PSS-SWCNTs nanocomposite films to various vapours of 10 ppm.

The chemical functionalization of CNTs can also help to improve the affinity for one specific analyte species over another during the sensing of analyte mixture. For example, chemically polymerized 3-methylthiophene in the presence of COOH-functionalized MWCNTs, mixed with polyethylene oxide (used as a binder) deposited between two palladium electrodes was sensitive to chloromethanes (CH_3Cl, CH_2Cl_2, $CHCl_3$, CCl_4, CH_4) with fast response times. However, it was not sensitive to acetone, acetaldehyde, benzaldehyde, tetrahydrofuran, methanol, and ethanol vapours providing high selectivity. The functionalized carbon nanotubes (CNT) with acidic groups (e.g.–COOH) have capability to doped PANI and can improve the sensing performances, for example good selectivity towards chloroform vapour over the other chlorinated methane vapour. The targeted functionalization of CNTs by conducting polymer via electrochemical route is demonstrated to facilitate creation of high-density individually addressable nanosensor arrays, with improved sensitivity, detection limit and reproducibility.

It is also shown that oxygen plasma treatment of CNTs and their alignment in the nanocomposites film can improve the sensing performance in terms of sensitivity, selectivity, rapidity of response, good reversibility and stability. Such effects are considered as manifestations of improved morphology and charge transport properties of thin film. For example, oxygen plasma functionalized and dielectrophoretically aligned CNTs-loaded PEDOT film-based sensors display excellent responses for detection of 2–300 ppm NH_3 and 6–1000 ppb trimethylamine gases at room temperature. The presence of CNTs is conjugated polymer matrix is also known to improve sensitivity due to increase in degree of interactions during adsorption or desorption of the analyte. For example, the sensitivity for NO_2 gas of polypyrrole—single-walled CNT nanocomposite is about ten times higher than that of pristine polypyrrole due to increase in the specific surface area by uniform polypyrrole coating on the single-wall CNTs. Nevertheless, though CNTs/polymer composites demonstrated to display improved sensing performance, the detailed sensing mechanism is still unclear, the cross-selectivity is poor and delayed recovery, stability and performance drift are the major issues to be resolved.

Applications of Polymer Nanocomposites

Biomedical Applications

The applicability of polymer nanotechnology and nanocomposites to emerging biomedical/biotechnological applications is a rapidly emerging area of development. One area of intense research involves electrospinning for producing bioresorbable nanofiber scaffolds for tissue engineering applications. This might be construed as a nanocomposite as the resultant scaffold allows for cell growth yielding a unique composite system. Another area also involving nanofibers is the utilization of electrically conducting nanofibers based on conjugated polymers for regeneration of nerve growth in a biological living system.

Nanoparticle silver, silver oxide and silver salts have been incorporated into polymer matrices to provide antimicrobial/biocidal activity. Nanoscale silver was shown to be a much better antimicrobial additive to nylon 6 than microscale silver particles due to the much higher rate of silver ion release. Nanosilver (5–50 nm) at levels of 0.1–1.0 wt% in poly(methyl methacrylate) bone cement exhibited high antibacterial activity for joint arthroplasty utility without the cytotoxicity of silver salts.

Polymer nanocomposites based on hydroxyapatite ($Ca_{10}(PO_4)_6(OH)_2$) have been investigated for bone repair and implantation. Hydroxyapatite, a major constituent of hard tissue, exhibits undesirable mechanical properties if directly employed thus polymer-based matrix composites are desired. Biodegradation of the matrix is also desired to allow infiltration of new bone growth at the repair site. Often natural polymers (polysaccharides, polypeptides, collagen, chitosan) or synthetic biodegradable polymers are employed as the matrix in these studies. Collagen derived gelatin and poly-2-hydroxyethylmethacrylate/poly(ε-caprolactone) nanocomposites based on hydroxyapatite are examples of systems studied for bone repair systems.

Electrospinning of biodegradable polymer solutions is a popular method to produce nanofiber scaffolds for tissue engineering applications. Poly(l-lactic acid)/exfoliated montmorillonite clay/salt solutions were electrospun followed by salt leaching/gas foaming. The resultant scaffold structure contained both nano- and micro-sized pores offering a combination of cell growth and blood vessel invasion micro-dimensions along with nanodimensions for nutrient and metabolic waste transport.

Polymer matrix nanocomposites have been proposed for drug delivery/release applications. The addition of nanoparticles can provide an impediment to drug release allowing slower and more controlled release, and reduced swelling and improved mechanical integrity of hydrogel-based nanocomposites. Iron oxide nanoparticles have been investigated for various applications including drug delivery, magnetic resonance imaging contrast enhancement, immunoassay and cellular therapy. These investigations often employ magnetite (Fe_3O_4) dispersed in a polymeric microsphere or microcapsule involving biodegradable and/or natural polymers. Poly(l-lysine) microspheres containing magnetic nanoparticles prepared by coacervation were prepared and characterized for potential use in targeted drug delivery applications. Iron and cobalt nanoparticles encapsulated in polydimethylsiloxane have been noted for treating retinal detachment disorders. A novel imprint lithography method based on a crosslinked perfluoroether templating mold has been employed to fabricate uniform micro- and nanoparticles from organic polymers including biodegradable polymers such as poly(lactic acid). Inclusion of inorganic nanoparticles has potential for a diverse range of biomedical applications using this method.

Fuel Cell Applications

Fuel cell applications involve polymers in the proton exchange membrane, binder for the electrodes and matrix for bipolar plates. The electrodes typically comprise carbon black particles (0.5–1.0 μm) with Pt catalyst particles of 2–5 nm and a polymeric binder (usually Nafion). Platinum nanoparticles deposited onto single-walled carbon nanotubes with Nafion as a binder were shown to give improved performance over the conventional carbon black-based electrodes. Nanoparticle incorporation in the proton exchange membrane has been noted in numerous publications to improve mechanical properties as well as to enhance proton conductivity. Additionally with direct methanol fuel cells, nanoparticles have been incorporated to reduce methanol crossover.

Heteropolyacids (HPA) (e.g., $H_3PW_{12}O_{40}$; $H_3PMo_{12}O_{40}$) have been added to proton exchange membranes to yield improved proton conductivity at higher temperatures while retaining good mechanical properties. The particle size of the HPA inclusions was generally in the nanorange. Silica nanoparticle inclusion in proton exchange membranes gave lower methanol crossover in several studies. Zirconium phosphate, zirconium hydrogen phosphate and TiO_2 nanoparticle incorporation in proton exchange membranes exhibited promise in direct methanol fuel cells. Nanoclay modified Nafion (laponite and montmorillonite) membranes have been reported to offer improvements over the unmodified controls. Sulfonated carbon nanofibers incorporated into a sulfonated EPDM proton exchange membrane yielded an order of magnitude improvement in the proton conductivity achieving performance comparable to the state-of-the-art Nafion 117.

Applications in Electronics Industry

- Nanotechnology is deeply embedded in the design of advanced devices for electronic and optoelectronic applications. The dimensional scale for electronic devices has now entered the nano range. The utility of polymer-based nanocomposites in these areas is quite diverse involving many potential applications and have been proposed for their use in various applications.
- They are used in chemical sensors, electroluminescent devices, electro catalysis, batteries, smart windows and memory devices.
- Another potential application includes photovoltaic (PV) cells and photodiodes, supercapacitors, printable conductors, light emitting diodes (LEDs) and field effect transistors.
- The electrical conductivity of carbon nanotubes in insulating polymers has also been a topic of considerable interest. The potential applications include electromagnetic interference shielding, transparent conductive coatings, electrostatic dissipation, super capacitors, electromechanical actuators and various electrode applications.
- Conjugated polymers with various nanoscale filler inclusions have been investigated for sensor applications including gas sensors, biosensors and chemical sensors. The nanofillers employed include metal oxide nanowires, carbon nanotubes, nanoscale gold, silver, nickel, copper, platinum and palladium particles.
- The emerging interest of graphene in electronics applications parallels that which occurred with carbon nanotube discovery. Graphene, in sheet form, may offer promise in replacing silicon as Moore's law limits are achieved.
- Polymer-based solar cells have the capability of being used to make cheap large flexible panels. The only downside is substantially low efficiency compared to commercial solar cells. Cadmium chalcogenides such as CdS nanoparticles in polymers have been used to make solar cells.
- Polymer semiconductor nanocomposites offer the promise of a new generation of hybrid materials with numerous possibilities of applications such as in optical displays, catalysis, photovoltaic, gas sensors, electrical devices, mechanics, photoconductors and superconductor devices.

References

- Schulz, Mark J.; Shanov, Vesselin N.; Yun, Yeoheung (2009). Nanomedicine Design of Particles, Sensors, Motors, Implants, Robots, and Devices. Artech House. ISBN 9781596932807
- Kumanek, Bogumiła; Janas, Dawid (May 2019). "Thermal conductivity of carbon nanotube networks: a review". Journal of Materials Science. 54 (10): 7397–7427. Bibcode:2019JMatS..54.7397K. doi:10.1007/s10853-019-03368-0. ISSN 0022-2461
- Carbon-nanotube-based-polymer-composites-synthesis-properties-and-applications, carbon-nanotubes- current-progress-of-their-polymer-composites: intechopen.com, Retrieved 1 June, 2020
- Fagan, Jeffrey (5 March 2009). "Carbon Nanotube Reference Materials". U.S. National Institute of Standards and Technology. Retrieved 6 September 2017
- "REACH Registration Completed for Single-Wall Carbon Nanotubes". pcimag.com. PCI Mag. 16 October 2016. Retrieved 24 November 2016

Ceramic Nanocomposites

3

- **Polymer–ceramic Nanocomposite**
- **Alumina/Chromium Carbide Ceramic Nanocomposite**

The nanocomposites that comprise of more than one solid phase in which at least one phase has dimensions in the range of 50nm to 100nm are called ceramic nanocomposites. A few of its examples are polymer-ceramic nanocomposite, alumina ceramic nanocomposite, chromium carbide ceramic nanocomposite, etc. This chapter discusses the subject of ceramic nanocomposites in detail.

Over the past half century ceramics have received significant attention as candidate materials for use as structural materials under conditions of high loading rates, high temperature, wear, and chemical attack that are too severe for metals. However, inherent brittleness of the ceramics has prevented their wide use in different applications.

Significant scientific effort has been directed towards making ceramics more flaw-tolerant through design of their microstructures by incorporation of fibers or whiskers which bridge the crack faces just behind the crack tip; by designing microstructures with elongated grains which act as bridges between crack faces just behind the crack tip; by incorporating second phase particles which deflect the crack making it travel a more tortuous path; and by incorporating secondary phases which undergo stress induced volume expansion that forces the crack faces together. However, one of the most recent development has been the distribution of multiple phases in a ceramic composite at the nanoscopic length scale. Owing to prevalence of nanoscopic features, such composites are referred to as ceramic nanocomposites.

The definition of nanocomposite material has broadened significantly to encompass a large variety of systems such as one-dimensional, two-dimensional, three-dimensional and amorphous materials, made of distinctly dissimilar components and mixed at the nanometer scale. The general class of nanocomposite organic/inorganic materials is a fast growing area of research. Reducing the sizes of structural features in materials leads to a significant increase in the portion of surface/interface atoms.

The surface/interface energies essentially control the properties of a solid. Interfaces provide a means to introduce non-homogeneity in the material. This non-homogeneity acts as a significant modification of both thermal and mechanical properties of the composites. Selective mixing of materials in a highly tailored morphology with high percentage of interface area, leads to materials with enhanced properties.

The properties of nano-composite materials depend not only on the properties of their individual parents but also on their morphology and interfacial characteristics. The nanocomposites find their use in various applications because of the improvements in the properties over the simpler structures. Few of such advantages can be summarized as:

- Improved mechanical properties e.g. Strength, modulus and dimensional stability.
- Decreased permeability to gases, water and hydrocarbons.
- Higher thermal stability and heat distortion temperature.
- Higher flame retardancy and reduced smoke emissions.
- Higher chemical resistance.
- Smoother surface appearance.
- Higher electrical conductivity.

For components used in a gas turbine engine, a lifetime upto 10000 h and a retained strength of ~300 MPa at a temperature of 1400 °C have been postulated, together with negligible creep rate. Furthermore, at elevated temperatures, the material must exhibit high resistance to thermal shock, oxidation, and subcritical crack growth. Ceramic nanocomposites have been shown to be extremely important for such future applications.

Advanced bulk ceramic composite materials that can withstand high temperatures (>1500 °C) without degradation or oxidation can also be used for applications such as structural parts of motor engines, catalytic heat exchangers, nuclear power plants, and combustion systems, besides their use in fossil energy conversion power plants. These hard, high-temperature stable, oxidation-resistant ceramic composites and coatings are also in demand for aircraft and spacecraft applications.

One such material system in this class of composites, Silicon Carbide/Silicon Nitride (SiC/Si_3N_4) composites, have been shown to perform very well under high temperature oxidizing conditions. Interest in such nanocomposites started with experiments of Niihara who reported large improvements in both the fracture toughness and the strength of materials by embedding nanometer range (20-300 nm) particles within a matrix of larger grains and at the grain boundaries. A 200% improvement in both strength and fracture toughness, better retention of strength at high temperatures, and better creep properties were observed.

An advanced nanocomposite microstructure such as that of polycrystalline Silicon Carbide (SiC)-Silicon Nitride (Si_3N_4) nanocomposites, figure, contains multiple length scales with grain boundary (GB) thickness of the order of 50 nm, SiC particle sizes of the order of 200-300 nm and Si_3N_4 grain sizes of the order of 0.8 to 1.5 μm. Designing the microstructure of such a composite (and similar others such as TiN-Si_3N_4, SiC-Al_2O_3, SiC-SiC, Graphene/CNT+SiC, and Carbon Fiber+SiC nanocomposites) for a targeted set of material properties is, therefore, a daunting task. Since the microstructure involves multiple length scales, multiscale analyses based material design is an appropriate approach for such a task.

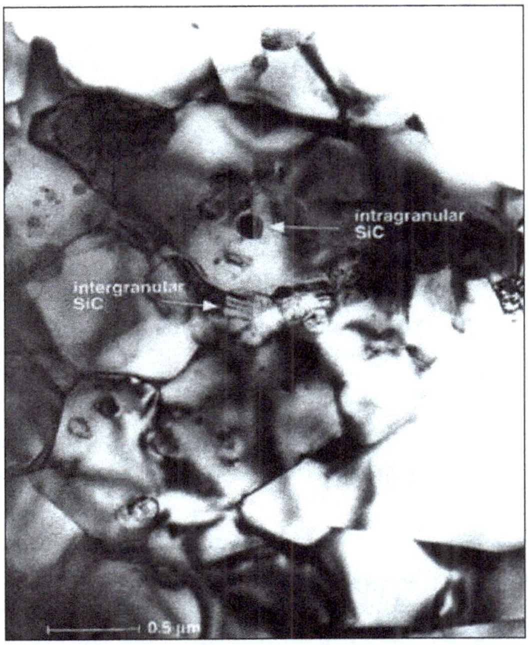

Actual microstructure of a SiC-Si_3N_4 nanocomposite.

The ceramic nanocomposite work in Multiphysics Lab at Purdue focuses on (1) Understanding Performance of Carbide and Nitride Based High Temperature Ceramic Nanocomposites for Extreme Environments found in power generation cycles Including Nuclear Applications, (2) Multiscale Modeling and Characterization in Oxide Ceramic Materials and (3) Understanding thermal conduction and thermal issues in materials for thermoelectric power generation. A description of major areas of interest and contributions is as follows:

- Understanding thermal conduction and thermal issues to develop materials with low thermal conductivity: This work focuses on understanding atomistic mechanisms of operation of nanocomposites for thermoelectric power generation such that materials with low thermal conductivity could be developed. Explicit molecular simulations using molecular dynamics (MD) are performed to understand how morphology alterations can be used to reduce thermal conductivity in nanocomposites. We have found certain biomimetic arrangements that could achieve significant reduction in thermal conduction. We are in the process of making and testing such materials.

- Understanding performance of carbide and nitride based high temperature ceramic nanocomposites for extreme environments including nuclear applications: This research work focuses on understanding mechanisms of room temperature and high temperature operations of advanced nanocomposite ceramic materials that can enable power plant operation at temperatures in excess of 1750 K leading to efficiencies of almost 70% and significant reduction in the plant emissions. As an off-shoot, this project also focuses on thermal properties of these materials for possible use as high temperature multifunctional materials, high temperature structural materials in nuclear applications or heat sensors in nuclear applications.

- Multiscale modeling and characterization in oxide ceramic materials: Focus during this work has been on understanding multiscale thermomechanical behavior of advanced composite materials such as multifunctional $Al^{+}Fe_2O_3$ nanocrystalline composites and high-strength Al_2O_3/TiB_2 ceramic armor composites. This research on atomistic deformation analyses of $Al^{+}Fe_2O_3$ multifunctional nanocomposites using MD is one of the first in the area of atomistic deformation analyses of advanced ceramic composite nanomaterials. In this work large scale MD simulations of nanocrystalline $Al^{+}Fe_2O_3$ multifunctional composites, of single crystalline Al, of single crystalline Fe_2O_3, and of various interfacial configurations of single crystalline Al and Fe_2O_3 are performed. In the case of Al_2O_3/TiB_2 ceramic armor composites, we have developed and used a new cohesive finite element method (CFEM) for quantitative characterization of dynamic fracture.

Multiscale Modeling of Ceramic Nanocomposites

Multiscale analyses (at nanometer and micrometer length and time scales) based on a combination of CFEM and MD based techniques have revealed that high strength and relatively small sized SiC particles act as stress concentration sites in Si_3N_4 matrix leading to inter-granular Si_3N_4 matrix cracking as a dominant failure mode. CFEM analyses have also revealed that due to a significant number of nano-sized SiC particles being present in micro-sized Si_3N_4 matrix, the SiC particles invariantly fall in wake regions of micro-cracks leading to significant mechanical strength. This finding was confirmed in the MD analyses that revealed that particle clustering along the GBs significantly increases the strength of these nanocomposites. While some nanocomposite morphologies have sharply defined SiC-Si_3N_4 interfaces, other nanocomposite morphologies have diffusion of C, N, or Si atoms at the interfaces.

In the case of SiC-Si_3N_4 nanocomposites, MD analyses have also revealed that the second phase particles act as significant stress raisers in the case of single crystalline Si_3N_4 phase matrix affecting the strength significantly. However, the particle's presence does not have a significant effect on the mechanical strength of bicrystalline or nanocrystalline Si_3N_4 phase matrices. The strength of the SiC-Si_3N_4 nanocomposite structures showed an uncharacteristic correlation between the grain boundary (GB) thickness and temperature.

The strength showed decrease with increase in temperature for structures having thick GBs having diffusion of C, N, or Si atoms. However, for structures with no appreciable GB thickness (no diffusion of C, N, or Si atoms), due to the particle clustering and increase in SiC-Si_3N_4 interfacial strength with temperature, the strength improved with increase in temperature. Figure shows snapshots of fracture propagation analyses in such nanocomposites obtained using the CFEM.

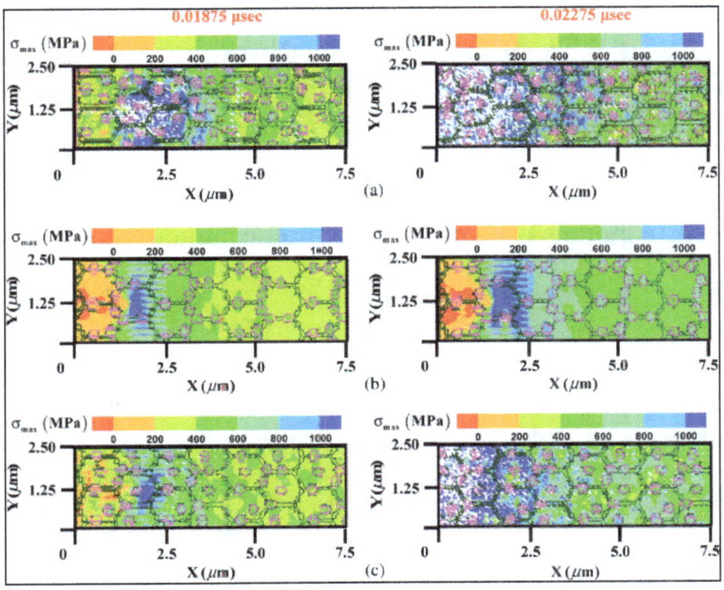

Snapshots of mesoscale crack propagation and damage propagation in the Si_3N_4 nanocomposites.

Figure displays snapshots obtained using MD. Current research work focuses on obtaining experimental images of the ceramic nanocomposites developed by collaborators, developing nanoscale CFEM meshes on such images, and performing failure analyses using the combination of MD and CFEM techniques.

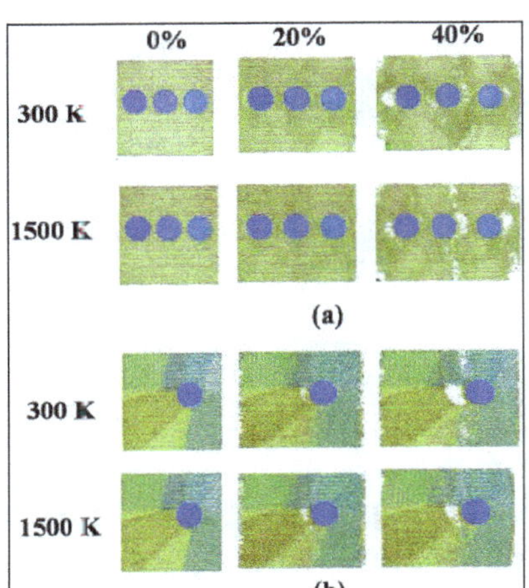

Snapshots of atomistic damage and failure propagation in two different SiC (particle) and Si_3N_4 (matrix) nanocomposites at two different temperatures.

Petascale Computing based Material Design

Atomistic analyses at the nanoscale can impart important information about the effect of critical features such as a GB, an interface, or a triple junction, etc. on mechanical deformation behavior of a small nanoscale (~ few nm) sample. In multiscale modeling such information is used to formulate macroscale (>few μm) material models for understanding microstructure dependent deformation behavior of a material sample.

Appropriate mathematical models of microstructure property relations allow to relate performances like fracture toughness, ultimate strength, fatigue lifetime etc., to key material microstructure parameters like volume fraction, particle size, and phase composition. Since a typical nanoscale test sample is much smaller and is subjected to varied surroundings in a typical microstructure, the incorporation of nanoscale information in macroscale models is subjected to statistical uncertainty.

If a complex microstructure is to be designed for a targeted set of properties, it is important that such uncertainties be correctly quantified and incorporated within a robust material design framework. We have pioneered the development of a variable fidelity model management framework that can incorporate material behavior analyses at multiple length scales in a design optimization framework.

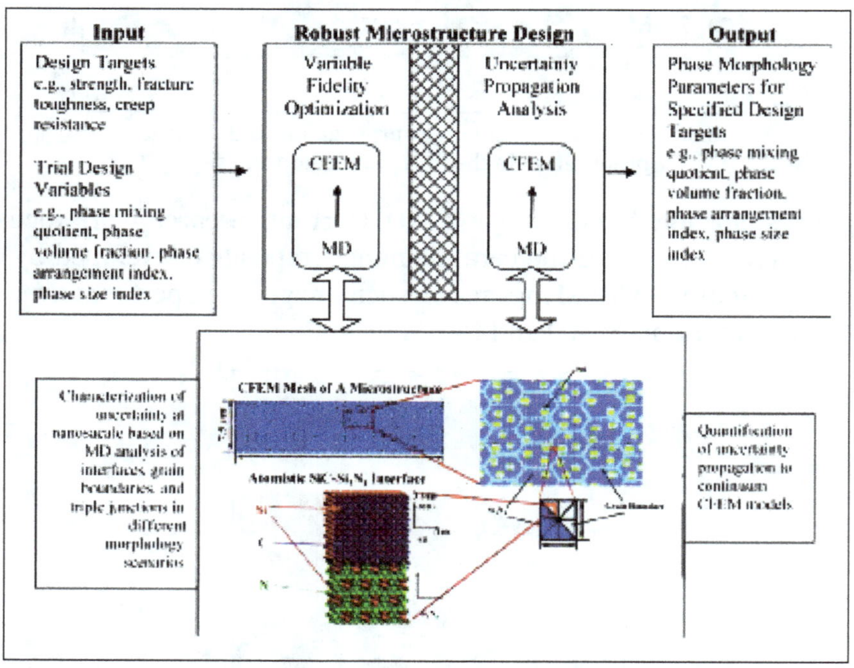

Schematic petascale material design framework.

Figure details the process flow of a petascale multi-physics model management tool for multiscale material design. Deployed on a petascale machine, the design tool developed that integrates atomistic and mesoscale analyses using a variable fidelity model management framework, will facilitate a significant reduction in nanomaterials' development cost and time with a simultaneous increase in the possible different combinations of individual composite material phases to achieve desired material performance.

The model management framework, besides managing the models and scales, is also well suited to control hierarchical parallelism. The natural hierarchy is MD within CFEM within design under uncertainty, using a mixed programming model SHMEM by SGI for CFEM and MPI for MD and the uncertainty modeling. Both MD and the uncertainty quantification (via quasi-Monte Carlo integration) can use 1000 processors, and CFEM 10, so 1000 uncertainty quantification groups of 10 CFEM groups of 1000 HMC processors is 10 processors, nearing exascale.

Preliminary material design analyses of the model system have been performed to understand the morphology related parameters that must be controlled for optimal targeted set of properties. The application of design tool is focusing on the continuous fiber ceramic composite (CFCCs) models of SiC-Si_3N_4 nanocomposites, The second phase (circles and cylinders) are the SiC fibers that have higher elastic modulus and higher creep resistance (E) but lower yield stress and fracture toughness, than that of the primary Si_3N_4 phase. The problem is to design the most suitable CFCC, with maximum strength and creep resistance for a set of external temperatures T, where the number of design variables will depend on whether the simulation tests are run on the 2-dimensional (2-D) or 3-dimensional (3-D) model. The design variables to be considered in the nanocomposite design optimization problem, for the 2-D model, are the fibers diameter (d) and the external temperature (T). And for the 3-D model the design variables to be considered are the fibers diameter (d), the length of fibers (l) and the external temperature (T). The problem definition in standard form is given below:

$$\left.\begin{array}{l}\text{minimize} f(d,l,T)=\{-\sigma_y(d,l,T),\dot{\varepsilon}_c(d,l,T)\} \\ \text{subject to}: d_{min}\le d\le d_{max} \quad l_{min}\le l\le l_{max} \text{ and } T=1500\,^{\circ}C\end{array}\right\}$$

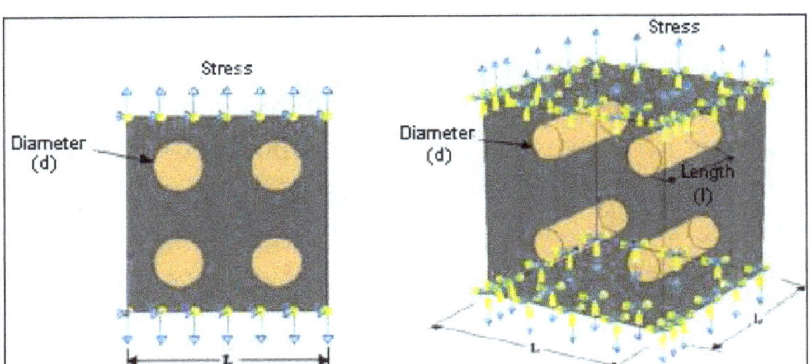

High and low fidelity models for the CFCC nanocomposites.

Figure illustrates normalized (0-100) function values for the strength and creep strain rate as a function of design variables for the high fidelity model (3-D) and low fidelity model (2-D). Figure (left) shows an increase in the CFCC strength and a corresponding decrease in the creep strain rate as the design variable d increases. Similarly for the high fidelity model, figure (right) shows an increase in the CFCC strength and a corresponding decrease in the creep strain rate as the design variables d and l increase.

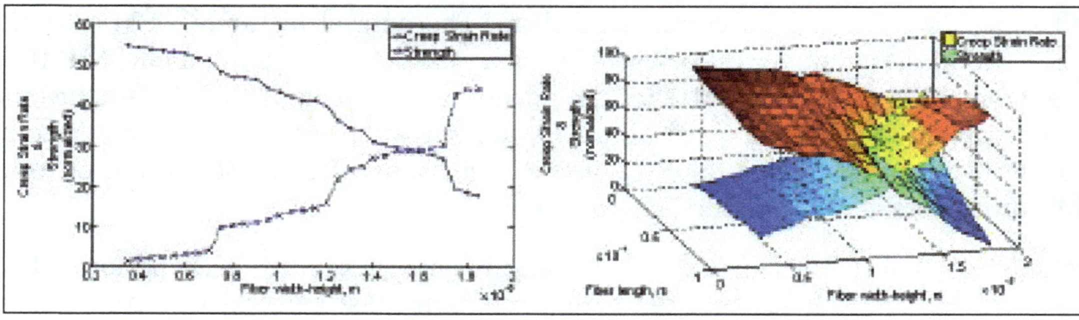

Figure (left) Strength and creep strain rate at 1500 °C as a function of the design variable width-height (d) for the 2-D low fidelity model. (right) Strength and creep strain rate at 1500 °C as a function of the design variables width-height (d) and length of fibers (l) for the 3-D high fidelity model.

Fabrication

Focus during this activity is on forming a collaborative modeling-deign-processing framework where complex ceramic nanocomposites for targeted set of mechanical and non-mechanical properties could be produced without wasting significant trial-and-error time and money. Polymer derived ceramics (PDCs) are an attractive approach to make material design predicted morphology of ceramic nanocomposites. First Niihara and his coworkers and then others used this approach to make high performance nanoscale reinforced composites.

Continued research in this area has led to the development of a range of nanostructures. One particularly interesting class of materials have predominantly amorphous Si-O-C nanodomains containing nanoscale SiC and C reinforcements. These materials have the desired characteristics for a broad range of high-temperature applications while offering greater control over processing, compositions and nanostructure. PDCs are produced by pyrolyzing preceramic polymers and are typically amorphous up to very high temperature but provide very intriguing ceramic-like properties, such as good creep and oxidation resistance.

Some of their unique properties are associated with in-situ formation of nanodomains and lack of grain boundaries in their microstructures. Due to the polymeric nature (thermoset) of the precursors, this family of materials is easily processable as fibers, matrices for composites, porous structures and coatings. Most studied PDCs can be categorized into three main groups: (i) Silicon carbide (SiC) (ii) silicon oxycarbides (SiOC) and (iii) silicon carbonitrides (SiCN). SiOCs and SiCNs are distinctive due to their hybrid molecular composition between SiO_2 and SiC and between SiC and Si_3N_4, respectively with additional level of "free" carbon as schematically illustrated in figure for Si-O-C system.

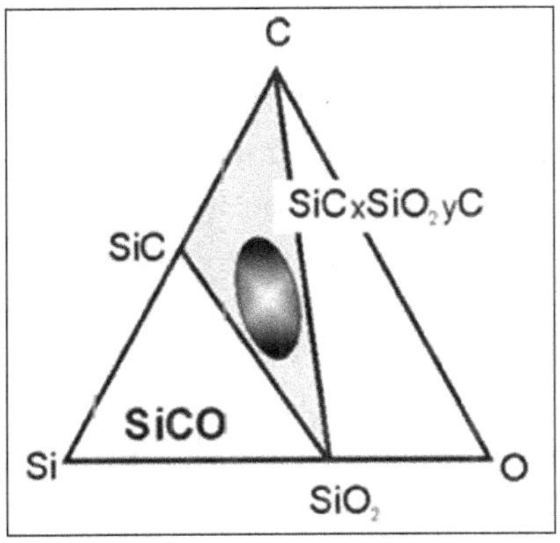

Schematic of phase relations in the Si-O-C system.

A unique nanostructural feature of these materials is that the controlled excess carbon is dispersed as graphene layers with domain size of a few nms. Control of, and understanding of development of such nanostructural features, using an integrated experimental and atomistic simulation approach.

Polymer–ceramic Nanocomposite

A polymer–ceramic nanocomposite is defined as an ionically conducting solid material derived from a polymer and a ceramic phase. This type of material is a subset of solid electrolytes.

A broader range of polymer and ceramic chemistries could be combined to synthesize polymer–ceramic nanocomposites. The incorporation of a ceramic phase into an ionically conducting matrix leads to enhanced conductivity, cationic transport number, and electrode–electrolyte interfacial stability. The ionic conductivity of the nanocomposite is dependent upon several variables such as the characteristics (chemistry, size, and volume fraction) of the ceramic phase, annealing parameters, physical and chemical properties of the polymer matrix, degree of reactivity between polymer and ceramic phases, and temperature.

A number of processing techniques (e.g. solvent casting, melt casting, and hot pressing) may be employed to obtain film and bulk nanocomposite specimens. The most convenient is the blend and hot press technique in which the polymer, salt (of lithium), and ceramic components are mixed in a predetermined proportion, ground in a mortar and pestle or energy milled, hot pressed into pellet form, and then rolled into films of desired thickness. Nanocomposite materials covering a wide range of concentration of ceramic phases in a polymer matrix can be obtained by this technique.

The temperature dependence of conductivity of the PEO:$LiBF_4$ complex and PEO:$LiBF_4$ (8:1)–MgO(10 wt%) nanomaterials containing micro- and nanosize MgO is shown in figure. The lowest conductivity values are associated with the PEO:$LiBF_4$ (8:1) complex. Near the melting point of PEO, 68 °C, a precipitous drop in conductivity begins, and at room temperature the conductivity drops to 10^{-9}S cm^{-1}. The specimen containing microsize (~5 micron) MgO exhibits much improved conductivity as compared to the PEO:$LiBF_4$ complex. At around ambient temperature, the conductivity is improved by approximately three orders of magnitude by the incorporation of microsize MgO in the polymer complex. The highest conductivity values are associated with the specimen containing nanosize MgO. The conductivity of this specimen is about four orders of magnitude higher than the PEO:$LiBF_4$ complex around the ambient temperature. Furthermore, the temperature dependence of conductivity diminishes as the MgO particles are reduced from micro- to nanosize. At 100 °C, all three specimens possess similar conductivity values, whereas the curves diverge as the temperature is lowered to 40 °C.

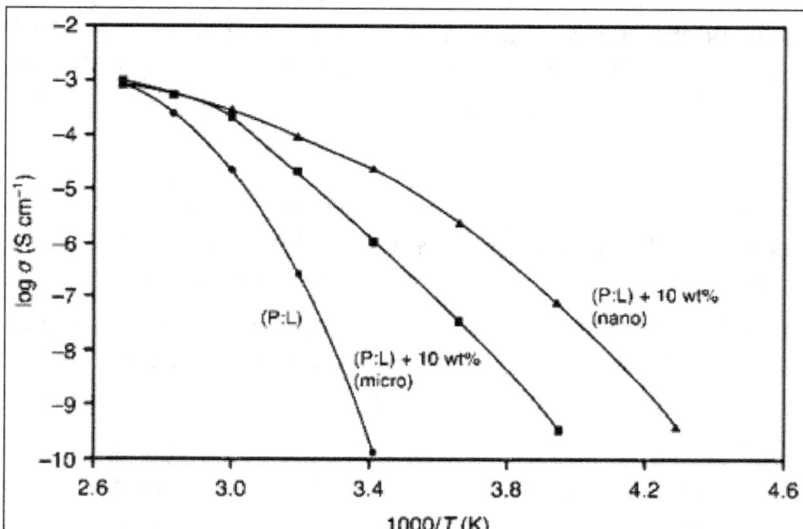

Temperature dependence of conductivity of PEO:$LiBF_4$ (P:L) complex (8:1), P:L + 10 wt% MgO(micro) and P:L + 10 wt% MgO(nano).

The temperature dependence of conductivity of the PEO:$LiBF_4$ complex and specimens containing micro- and nanosize $BaTiO_3$ is shown in figure. These specimens also exhibit behavior similar to the MgO-containing electrolytes shown in figure. The low-temperature conductivities are greatly improved by the incorporation of $BaTiO_3$. The nano- and microsize $BaTiO_3$ specimens, however, exhibit a crossover at around 50 °C. The nanosize specimen possesses lower conductivity above 50 °C but higher conductivity below 50 °C as compared to the microsize $BaTiO_3$ specimen. This trend was also noted for MgO in figure, but to a much lower degree. The ambient temperature conductivities of $BaTiO_3$-containing specimens are in the range 10^{-4}–10^{-5} S cm^{-1}.

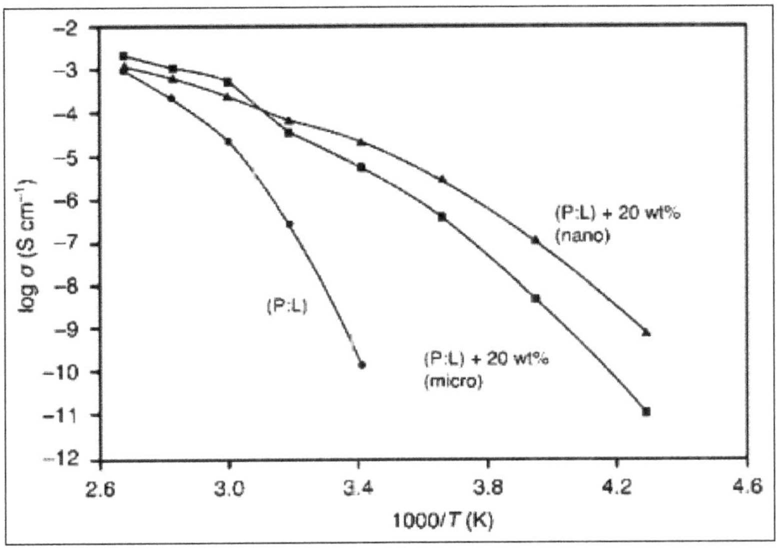

Temperature dependence of conductivity of P:L + 20 wt% $BaTiO_3$(micro) and P:L + 20 wt% $BaTiO_3$(nano).

Table presents the physical properties of MgO and $BaTiO_3$ materials. $BaTiO_3$ is an important ferroelectric material and its dielectric constant in the tetragonal phase (ferroelectric state) is much higher and anisotropic. The dielectric constant of $BaTiO_3$ is also known to be particle-size dependent. The two ceramic materials, MgO and $BaTiO_3$, in effect represent the upper and lower limits of dielectric constants of available ceramic materials. The last column of table shows masses corresponding to the two particle sizes used for each material. For MgO and $BaTiO_3$, the masses corresponding to each particle size differ by approximately seven and three orders of magnitude, respectively.

Table: Physical properties of MgO and $BaTiO_3$.

Ceramic oxide	Melting temperature (°C)	Density (g cm^{-3})	Dielectric constant (k)	Mean particle size	Mass of particle (g)
MgO	2852	3.58	9.65	5 µm	2.34×10^{-10}
				20 nm	1.50×10^{-17}
$BaTiO_3$	—	6.02	$\kappa_{11} = 3600$	1 µm	9.45×10^{-12}
		(tetragonal)	$\kappa_{33} = 150$	70 nm	8.65×10^{-15}

In spite of widely different physical characteristics, including density and dielectric constant, both MgO and $BaTiO_3$ additives exhibit similar effects with regard to particle size, and thus this is believed to be a dominant parameter. Reducing the particle size from the micro- to nano-range increases low-temperature conductivities and decreases their temperature dependence, as shown in figure. The enhancements are pronounced and associated with lower activation energies for the ionic transport.

The average length and mass of a PEO chain are 20 µm and 3×10^{-18}g, respectively. These length and mass values must now be compared with the particle size and mass of the ceramic dopants to develop insight into possible interactions. The proximity and interaction of a polymer chain and a nanosize MgO particle (0.02 μm) are shown schematically in figure. A constant segmental chain motion near the glass transition temperature is expected to cause displacement and distribution of MgO particles. Furthermore, the process can be facilitated by thermal treatments and cycling. The displacement and distribution of MgO particles continues until the polymer chain dipole and MgO dipole interaction takes place. The interaction leads to latching of the polymer chain and MgO particle, and thus stabilization of the structure. When experimentally monitored, this phenomenon is reflected by an increase in conductivity during isothermal stabilization at low temperatures.

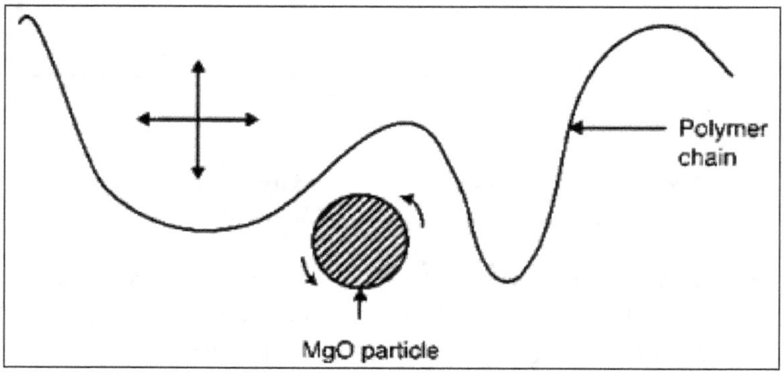

Schematic representation of a polymer chain segment and MgO particle interaction.

The degree of polymer chain and ceramic particle interaction can be reduced by a heavier, fewer, and lower weight or volume percent of MgO. For example, if the weight percent of MgO is maintained constant and the particle size is reduced from 5 μm to 20 nm, over 15 million additional polymer-ceramic interaction sites are created and conductivity enhancements occur. These lighter MgO particles become far more receptive to the segmental chain motion of the polymer, leading to improved polymer chain and MgO particle interaction. Such an interaction is reflected by an enhanced and time-dependent conductivity.

It is imperative that conductivity is directly related to the mobility and concentration of ions. Because the concentration is maintained constant in a given specimen, the mobility must increase to account for the conductivity enhancement. It has been suggested that the mobility of charge carriers in nanocomposites is increased by the annealing-induced structural changes that occur because of an interaction between polymer chains and ceramic particles.

The fact that motions of polymer chains contribute to the transport of ions in polymer electrolytes also has deleterious effects on the transport number. The chain motion also facilitates transport of larger anionic species and thus measured conductivity includes a contribution from both cationic and anionic species. Lithium in PEO-based electrolytes is coordinated with five oxygen atoms, and the probability of finding a similar site for lithium transport is low. The polymer electrolytes have a lithium ion transport number that may be as small as 0.3. As a result, when such materials are used in a lithium cell, extensive concentration gradients are set up during use and these affect their performance.

The cationic transport number in amorphous inorganic and polymeric electrolytes can be approximated through the use of the decoupling constant, R. The larger the value of R, the greater is the structural relaxation time and less is the transport of ions (both anions and cations) mediated by the structure. In polymer–ceramic composite electrolytes, the glass transition temperature, T_g, is proportional to the volume fraction of the ceramic phase. Angell has shown that for an R value of ~10^2, the cationic transport number could be over 0.9, as the transport of ions mediated by the structure would diminish. An increase of 50 °C in the T_g in most polymer electrolytes resulting from the addition of ceramic phase will bring the T/T_g ratio to 1.2 ($R \simeq 10^2$) and the cationic transport number to around 0.9.

The conductivity of a polymer electrolyte originates from two distinct processes: ion hopping and ion transport assisted by polymer chain motion. The addition of a ceramic phase suppresses the chain motion-mediated contribution and thus must increase the contribution associated with ion hopping if the conductivity remains the same (the worst case scenario). The ion hopping process is more favorable for cationic species than anionic species because of their small size and mass. This scenario suggests an enhanced cationic transport number as the volume fraction of ceramic phase increases in the polymer matrix of the polymer–ceramic nanocomposites.

The conductivity and transport number of nanocomposities comprising LiI, PEO, SiO_2, MgO, and Al_2O_3 have been measured and reported by Nagasubramanian et al. and Peled et al. They calculated the conductivity from bulk resistance, R_b, measured at high frequency and transport number, t^+, using:

$$t^+ = R_b / (R_b + Z_d)$$

where Z_d is diffusional impedance as measured from a Nyquist plot. For a composite electrolyte film containing 0.05 μm alumina, the bulk conductivity is around 10^{-4} Scm^{-1} and the lithium ion transport number is close to unity at 104 °C.

Polymer–ceramic nanocomposites show properties intermediate between organic polymers and inorganic ceramers. The primary goal of preparing such composites is to achieve combinations of properties from both components.

However, due to the incompatibility of two dissimilar phases involved, undesirable phase separation may often result. Hence the preparation of nanocomposites with homogeneous dispersion has been a critical issue. The interaction between the organic phase and the inorganic phase is paramount in constraining phase separation. The possible interactions between the organic phase and the inorganic phase can be achieved through hydrogen bonding, binding the dissimilar phases with the direct use of a bonding agent or by surface modifications of the organic polymer or the inorganic ceramic.

Improved interfacial bonding exerts a profound effect on the morphology of the resulting hybrid materials and it tends to suppress agglomeration of ceramic particles, resulting in reduced particle sizes, thus increasing the surface area which can interact with the polymer. This allows the ceramic particles to interact with the polymer backbone at a molecular level, resulting in substantial reinforcement of the hybrid material. Enhancements in interfacial bonding in various polymer–ceramic nanocomposites giving special emphasis on high-temperature polymer–ceramic nanocomposites are addressed.

Polymer–ceramic nanocomposites offer significant potential advantages in biomedical applications where the properties of either polymers or ceramics alone are insufficient to meet performance requirements. The use of nanotechnology in medicine offers some exciting possibilities. Some techniques are only imagined while others are at various stages of testing, or actually being used today. Nanotechnology in medicine involves applications of nanoparticles currently under development as well as longer-range research that involves the use of manufactured nanorobots to make repairs at the cellular level. The use of nanotechnology in the field of medicine could revolutionize the way we detect and treat damage to the human body and disease in the future, and many techniques only imagined a few years ago are making remarkable progress toward becoming realities.

Alumina/Chromium Carbide Ceramic Nanocomposite

Alumina is one of the important materials in structural ceramic application because of its excellent mechanical properties, good chemical stability and high temperature characteristics. Its intrinsic brittleness and relatively poor reliability however, made the toughening of alumina ceramics an important and challenging area of research now days. The incorporation of secondary ceramic reinforcement phases (e.g. particulates, fibers, or platelets) has been proven to be an easy, safe and economically toughening technique for alumina ceramics. There are several approaches by the researchers to toughen the alumina matrix with addition of different secondary phases. Among all the approaches, SiC reinforcement particles on Al_2O_3 composites mostly improve the mechanical properties. But the toughness enhancement is not significant for these composites. On the other hand, incorporating chromium carbide particles into Al_2O_3 matrix has been successfully achieved for toughening purposes. These aluminum oxide/chromium carbide composites demonstrate superior mechanical properties to those of matrix phase due to the high Young's modulus of chromium carbide and its outstanding ability to resist high temperature erosion at temperatures up to 100 °C. In addition, the good electrical conductivity of chromium carbide makes electrical discharge machining possible.

Besides second phase particles, solid solution strengthening is another mechanism for the Al_2O_3 matrix. Chromia (Cr_2O_3) has been used to improve the physical properties of Al_2O_3. As Cr_2O_3 has the same corundum crystal structure similar to Al_2O_3, Al_2O_3–Cr_2O_3 can form substitutional solid solution in all ranges at high temperature. The addition of Cr_2O_3 was found to increase the hardness, tensile strength, and thermal shock resistance of Al_2O_3.

The flow chart showing the detailed experimental procedure is given in figure.

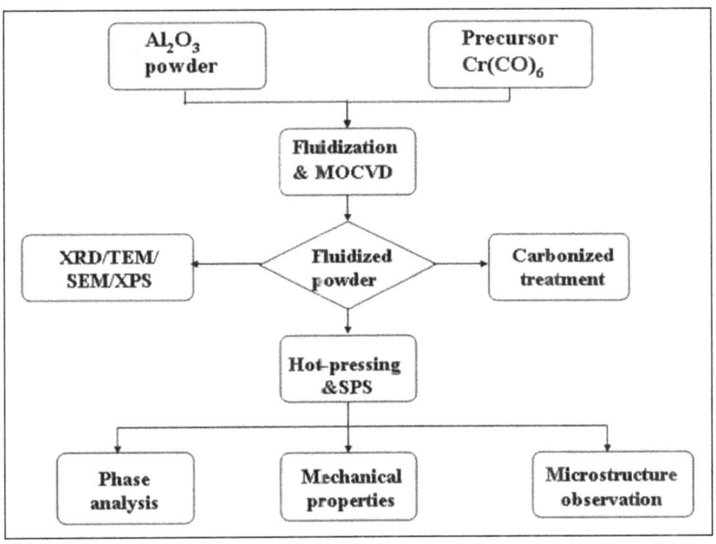

Flow chart of the experimental procedure adopted in the current work.

Powder and Sample Preparation

Chromium hexacarbonyl (Cr(CO)6, 99%, Strem Chemicals Co., USA) was used as the precursor of chromium oxide in the MOCVD process. Aluminum oxide powder with the average particle size of about 0.2–0.4 µm (A16-SG, Alcoa, USA) was used as the matrix powder. A vaporized precursor carried by He gas (99.9% pure) was introduced into the fluidized bed reactor for the MOCVD process. Based on the results of Lander & Germer, 1947, the precursor container was kept at 75 °C in a vacuum (10 torr) in the present experiment. To investigate the influence of the temperature on the product deposited on the fluidizing Al_2O_3 particles, the Cr(CO)6 vapor was decomposed in the fluidized chamber at 300 and 400 °C, respectively. The apparatus consisted of six main components: gas supply, MOCVD precursor, fluidized bed reactor, power supply, temperature controller, cold trap and vacuum system, as shown schematically in figure. The pressure in the chamber was measured with a pressure meter. A rotary vacuum pump was set up with a cold trapping system. The temperature controllers (TC) were connected to a heating system that controlled the temperature of the fluidized reactor and precursor container.

Characterization of the Fluidized Powder

The fluidized powders were analyzed by x-ray diffraction (XRD; Rigaku D/MaxII, Japan), and transmission electron microscopy (TEM; Hitachi FE-2000, Japan) equipped with energy dispersive x-ray spectroscopy (EDS; UK). X-ray photoelectron spectroscopy (XPS; VG Scientific 210, UK) was used to determine the coating phases by binding energy. A Brunauer–Emmett–Teller (BET) instrument Micromeritics Gemini 2360, USA) was used to measure the specific surface area by nitrogen adsorption, and C/O analyzer (LECO CS-244, USA) was used to measure the carbon content of coated powder by a combustion method. Figure shows the XRD patterns of decomposed Cr $(CO)_6$ prepared in fluidized bed at 300 °C and 400 °C. It is observed that at 300 °C, the powder is completely amorphous in nature, where as at 400 °C, the powder consists of Cr_2O_3 along with CrC_{1-x} phase.

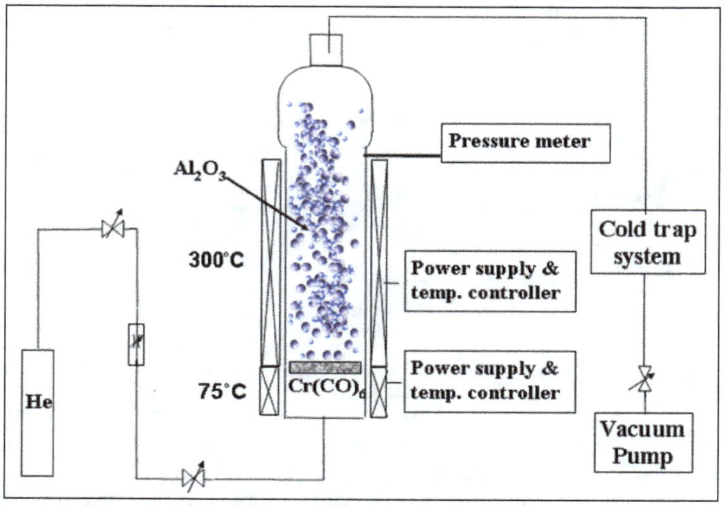

Schematic diagram of MOCVD and fluidized bed.

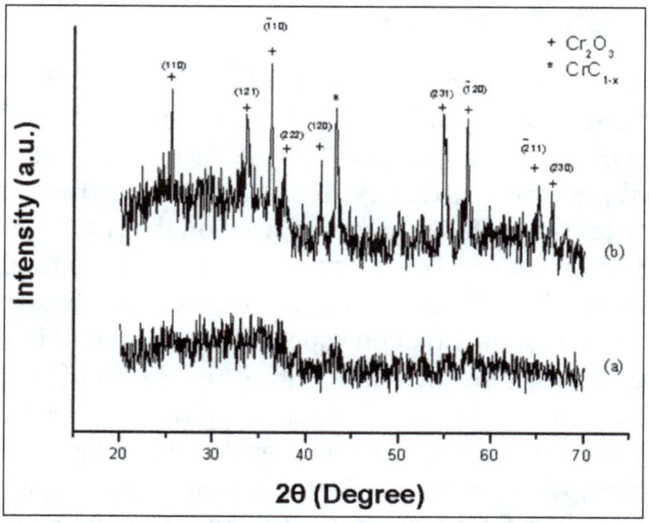

XRD pattern of decomposed Cr (CO)6 prepared in fluidized bed at (a) 300 °C (b) 400 °C.

The pyrolysis of precursor $Cr(CO)_6$ at 300 °C results the formation of Cr_2O_3, C-Cr and C-C bonds observed from XPS spectra. This figure displays the XPS spectra of the Cr 2p and C 1s regions of the as-deposited powder prepared at 300 °C. Figure shows two peaks corresponding to the spin-orbit splitting 2p1/2 (right) and 2p3/2 (left) of Cr with bonding energies of 586.3 and 576.6 eV, respectively. The band shift of these two peaks is 9.7 eV, which is in good agreement with previously reported data and confirms the existence of Cr_2O_3 particles in the as-deposited powder. The XPS spectra of the C 1s regions in figure. provide the evidence for which at least two forms of carbon in the as-deposited powder exist. One is free carbon (C–C) at 284.6 eV, and the other carbon bonded to chromium atoms (C–Cr) at 283.5 eV.

The decomposed precursor deposits uniformly over Al_2O_3 and the deposited nano particles (~ 30 nm) are shown in the TEM micrograph. The insert shows the TEM pattern of the deposited nano particles, which is amorphous in nature. From the EDS spectrum shown in figure, it is confirmed that the nanoparticles consist of mostly Cr and C along with small amount of C. The presence of Cu is the contribution from Cu grid.

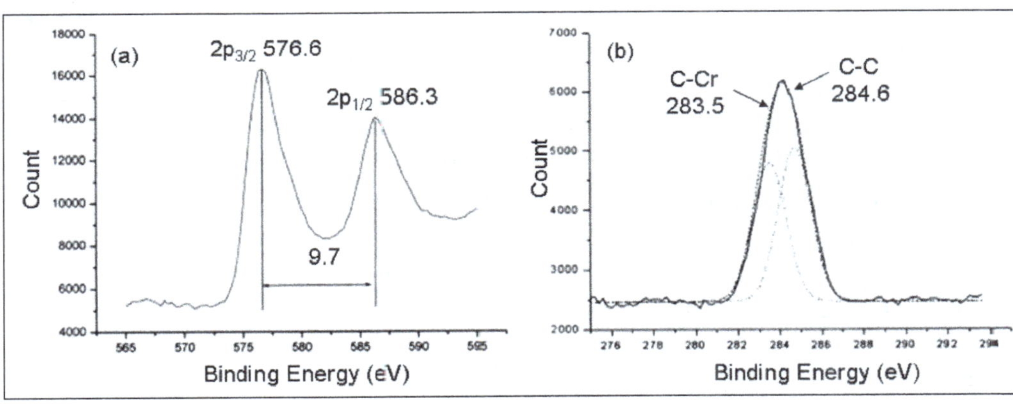

X-ray photoelectron spectra of the (a) Cr 2p regions and (b) C 1s regions of as deposited composite powders prepared at 300 °C in fluidized bed.

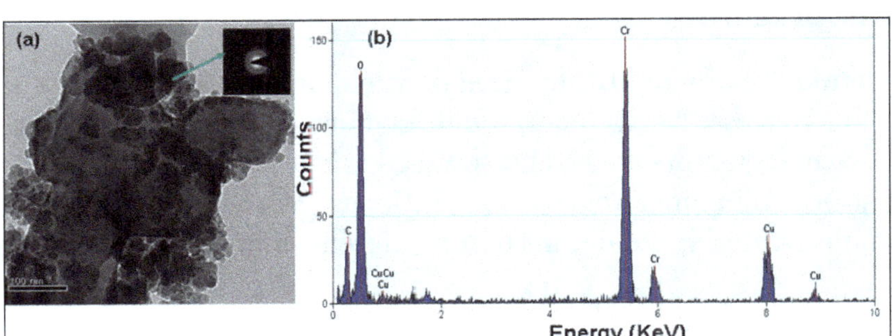

TEM micrographs of (a) nanoparticles deposited on alumina particle (b) EDS spectrum of the coating particle.

Table shows the carbon content of the samples prepared at different fluidized temperature. The carbon contain is 0.75 and 0.23% for the samples fluidized at 300 and 400 °C, respectively. It indicates that the as-deposited amorphous powder prepared at

300 °C has more carbon content than the crystalline powders prepared at 400 °C. This was probably because of the amorphous powder has specific surface area (25.5m²/g) larger than the crystalline powder (11.2 m²/g).

Table: Specific surface area and carbon content of specimens.

Specimen	Fabricated in Fluidized reactor	Carbon content (%)	BET surface area (m²/g)
1	300 C, 2h	0.75	25.5
2	400 C, 2h	0.23	11.2

Carbothermal Treatment of Cr_2O_3 Particles Dispersed on Al_2O_3 Particles

The carbothermal treatment of the prepared Cr_2O_3/Al_2O_3 powders was first performed in a graphite furnace at a variety of temperatures from 700 to 1150 °C in a vacuum condition (10^{-3} Torr) for 2h. XRD patterns of the fluidizing powders treated from 700 to 1150 ◦C in the graphite furnace for 2 h. The Cr_2O_3 peaks seen in the XRD patterns clearly indicate that the amorphous Cr_2O_3 has crystallized, but does not carbonize at temperatures under 1000 °C. As reported by Bouzy et al. and Bewilogua et al. an annealing treatment causes a transformation of the metastable carbide CrC_{1-x} into the stable carbide phase Cr_3C_2. Here, Cr_3C_2 peaks are not observed in the XRD patterns, consequently, suggesting that the CrC_{1-x} content of the decomposed Cr $(CO)_6$ is too little to be found in XRD patterns. According to the XRD pattern shown in figure, Cr_2O_3 reacts with carbon and transforms into Cr_3C_2 when the treatment temperature is 1150 °C. This carbothermal reaction process can be shown as the following equation:

$$Cr_2O_3(s) + 13/3C(s) \rightarrow 2/3Cr_3C_2(s) + 3CO(g)$$

The Formations of Cr-carbide/Al_2O_3 Nanocomposite and Cr_2O_3/Al_2O_3 Solid Solution

After carbothermal treatment, the fluidized powders undergone hot pressed sintering at 1400 °C. For this process, the 200 mesh fluidized powder was first die-pressed, and then put into a BN-coated graphite die and hot-pressed at a pressure of 25MPa in a HP furnace (High-multi 5000, Fujidempa Kogyo Co., Ltd., Japan) at 1400 °C under vacuum (5×10^{-4} torr). The detailed sintering conditions of the different samples are listed in table.

Table: Sintering conditions of hot pressed samples.

Samples	Sintering powders	Sintering conditions
ALO	Pure alumina powder	1400C/1h, pressure 25 MPa
Sample 1	Composite powders fluidized at 300 C for 1 h	1000C/1h, 1400C/1h, 25 MPa
Sample 2	-do-	1150C/1h, 1400C/1h, 25 MPa
Sample 3	-do-	1400C/1h, pressure 25 MPa

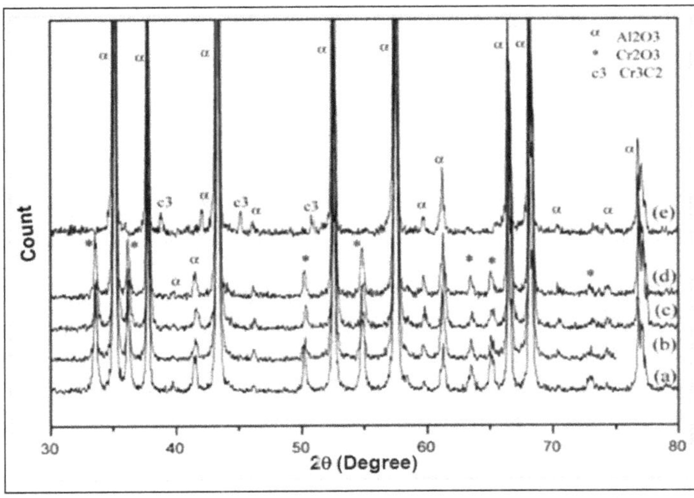

XRD patterns of the powder thermally treated at a variety of temperatures: (a) 700 °C, (b) 800 °C, (c) 900 °C, (d) 1000 °C, and (e) 1150 °C in a vacuum in a graphite furnace for 2 h for the fluidized powders prepared in the fluidized bed.

Figure shows the XRD patterns of the samples thermally treated under different conditions. The Si peaks shown in the XRD pattern were used for calibration. First, figure indicate that the Al_2O_3 peaks for sample 1 shifted to lower angles than those for ALO. For pure ALO, the pure alumina powder was hot pressed at 1400 °C, while for sample 1, the fluidized powder was pre-sintered at 1000 °C before the hot pressing at 1400 °C. During the pre-sintering at 1000 °C, most of the Cr_2O_3 reacted with Al_2O_3 to form an Al_2O_3–Cr_2O_3 Solid solution and thus the Al_2O_3 peaks shifted to lower angles in the XRD pattern. Furthermore, the color of the pieces is dark red as observed in sample 1, due to the (Al, Cr)$_2O_3$ solid solution reaction product. Secondly, a comparison with the peaks of ALO reveals that the Al_2O_3 diffraction peaks of sample 2, as shown in figure, shift to lower angles, occurring simultaneously with the peaks of Cr_3C_2. Consequently, besides forming a solid solution, some of the Cr_2O_3 reacts with carbon as equation (Eq. 1) to form chromium carbide Cr_3C_2 when the powders have been pre-sintered at 1150 °C. Finally, figure shows the XRD patterns of sample 3, which is similar to the results of sample 2. In addition to forming a solid solution, some of the Cr_2O_3 was also carbonized. However, it transformed into mixed phases of Cr_3C_2 and Cr_7C_3 when the fluidized powders were hot pressed at 1400 °C. Berger et al., 2001 reports that Cr_7C_3 is formed at an elevated temperature (>1150 °C). Moreover, the TG/DTA results of our paper also indicate that the generation temperature of Cr_7C_3 is about 1170 °C. This carbothermal reaction process can be shown as reaction. Consequently, not only Cr_3C_2, but also Cr_7C_3 is formed in sample 3, hot pressed at 1400 °C.

$$Cr_2O_3(s) + 27/7C(s) \rightarrow 2/7Cr_7C_3(s) + 3CO\ (g)$$

Figure shows the SEM micrograph of pure Al_2O_3 and 8b is the High Angle Annular Dark Field scanning transmission electron microscopy (HAADF STEM) image of sample 1, showing almost all the Cr^{3+} replaced the Al^{3+} and formed an Al_2O_3–Cr_2O_3 solid solution.

According to Bondioli et al., 2000, at temperatures of over 1000 °C the complete ranges of substitutional solid solutions are obtained. Figure are the HAADF STEM images of samples 2 and 3, respectively, showing that in addition to forming a solid solution, nanosized chromium carbide particles also disperse uniformly in the alumina matrix. Comparing the HAADF STEM images of samples, sample 2 has smaller alumina grain size (0.7m) than others (sample 1: 1.5m, sample 3: 0.9m.) This is because it has much nanosized particles on the Al_2O_3 grain boundaries, which effectively inhibit the grain growth. The volume percentages of the reinforced particles showing in figure are 4.5 and 1.5 vol%, respectively. Conversely, the Al_2O_3 grains of sample 1 have a larger growth size as a result of the diffusion of Cr atoms inducing grain boundary migration (DIGM). The drive force for the migration is believed to rise from the coherency strain in the Cr_2O_3 diffusion zone in front of the migrating grain boundaries.

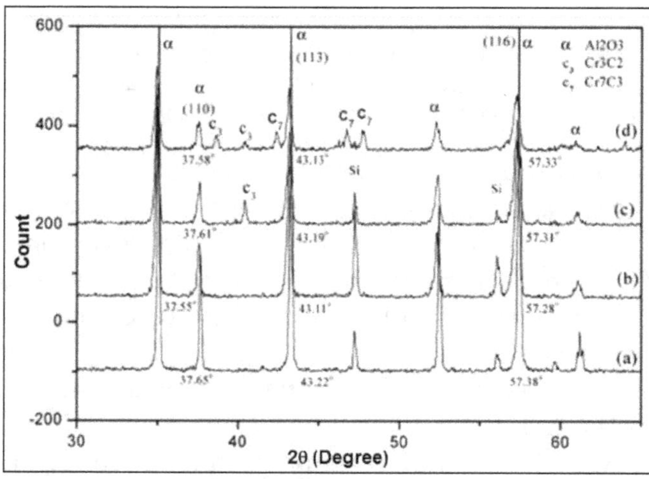

XRD patterns of hot pressed samples: (a) ALO, (b) sample 1, (c) sample 2, and (d) sample 3.

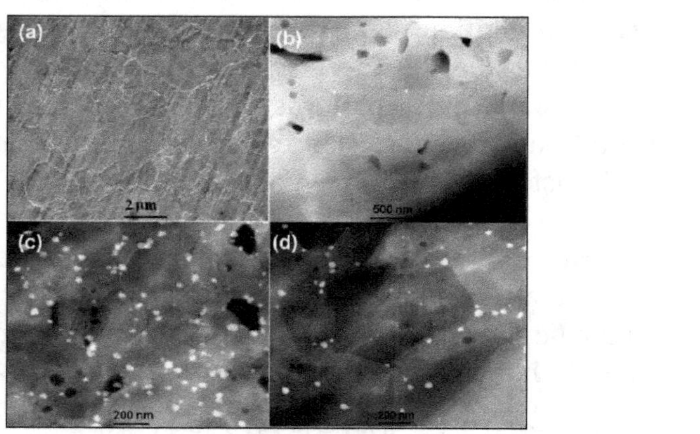

SEM micrograph of (a) pure Al_2O_3; HAADF STEM micrographs of samples: (b) sample 1: Al_2O_3-Cr_2O_3, (c) sample 2: Al_2O_3- Cr_3C_2 and (d) sample 3: Al_2O_3- Cr_7C_3.

There are two HRTEM micrographs of interfaces shown in figure One is the interface between the Cr_3C_2 and Al_2O_3 and the other is the interface between Cr_7C_3 and Al_2O_3. The interface between Al_2O_3 and Cr_3C_2 is non-coherent. The smaller illustrations in figure show Al_2O_3 with hexagonal and Cr_3C_2 with orthorhombic structure. The lattice spacing of the Al_2O_3 ($\bar{1}$10 1 4) is 0.255 nm and that of the Cr_3C_2 (0 0 2) is 0.115 nm. The spacing difference between these two planes is large. However, the interface between Al_2O_3 and Cr_7C_3 is semi-coherent. Structures of both Al_2O_3 and Cr_7C_3 are hexagonal. The lattice spacing of the Al_2O_3 plane ($\bar{1}$ 1 0 2) and Cr_7C_3 plane (2 $\bar{2}$ 0 1) are 0.349 and 0.360 nm, respectively. There is little difference between the lattice spacing of these two hexagonal planes.

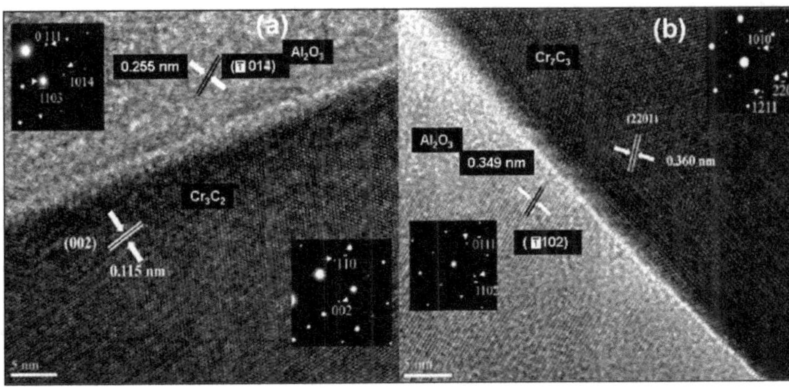

Lattice images of the interfaces (a) between Cr_3C_2 and Al_2O_3, and (b) between Cr_7C_3 and Al_2O_3.

Carbothermal Reduction of Cr_2O_3 using Methane as Carbonization Source

The conventional route for Cr_3C_2 composite synthesis mainly involves the reduction of chromium oxide by methane gas. However, besides the three stable chromium carbides, Cr_3C_2, Cr_7C_3, and $Cr_{23}C_6$, several metastable carbides such as Cr_3C, CrC_{1-y}, Cr_5C_2 and Cr_3C_{2-x} also have been reported, which emanate during different preparation methods and for different Cr/C contents. Lerch & Rousset indicated that the $CrO_{1.9}$ with high surface area (200–350 m²/g) reduced at 700 °C in an atmosphere of methane and hydrogen mixture leading to the formation of stable Cr_3C_2 and metastable Cr_3C_{2-x}. The metastable Cr_3C_{2-x} powder was also prepared by Loubière et al. using CH_4–H_2 atmosphere from the carburization of metastable chromium oxide. The metastable Cr_3C_{2-x} is a Re_3B-type structure, readily formed by carburization of high specific surface area chromium oxide and usually co-existent with Cr_2O_3, Cr_3C_2 and free carbon.

We have discussed the synthesis of Cr_2O_3/Al_2O_3 composite powders by the decomposition of chromium hexacarbonyl in a fluidized bed. This was followed by the hot-pressing of the composite powders in a graphite mode eventually leading to its transformation into a Cr_3C_2/Al_2O_3 nanocomposite. One of the major drawbacks in this synthesis process was the formation of undesirable Al_2O_3 - Cr_2O_3 solid solution at a high temperature, which leads to abnormal grain growth of the Al_2O_3 matrix and eventually decreased the desired Cr_3C_2 contents in the composite. In order to carburize the chromium oxide completely, a new approach is reported by using the carburization process of nanosized chromium oxide formed from a metal-organic chromium hexacarbonyl precursor in a carburizing mixture of CH_4/H_2 gas. For this process, the as received powder after fluidization was carburized in an alumina tube furnace in the presence of a methane–hydrogen mixture (1:9) at 700–850 °C for 5 h. The H_2 restrain the CH_4 to decompose fast and using CH_4 as carbon source, the carbothermal reduction reaction becomes:

$$3Cr_2O_3(s) + 13CH_4(g) \rightarrow 2Cr_3C_2(s) + 9CO(g) + 26H_2$$

Where, $\Delta G = 262564.44 - 236.79T$ (Cal) and ($T > 841$ °C)

Figure shows the XRD patterns of the decomposed powder carburized at 700–850 °C for 5 h in CH_4–H_2 gas. At the carburization temperature of 700 °C, the pattern displays sharp peaks that index with the Cr_2O_3 phase. At the increased carburization temperature of 750 °C, the intensity of Cr_2O_3 peaks decreases significantly accompanied by some incipient low intensity peaks which could be indexed with metastable Cr_3C_{2-x}. When the temperature reaches to 800 °C, the Cr_2O_3 phase disappears, concomitant with a complete transformation to the Cr_3C_{2-x} phase. Further increase of the carburization temperature to 850 °C, leads to the formation of the stable Cr_3C_2 phase along with a small amount of Cr_3C_{2-x} metastable phase.

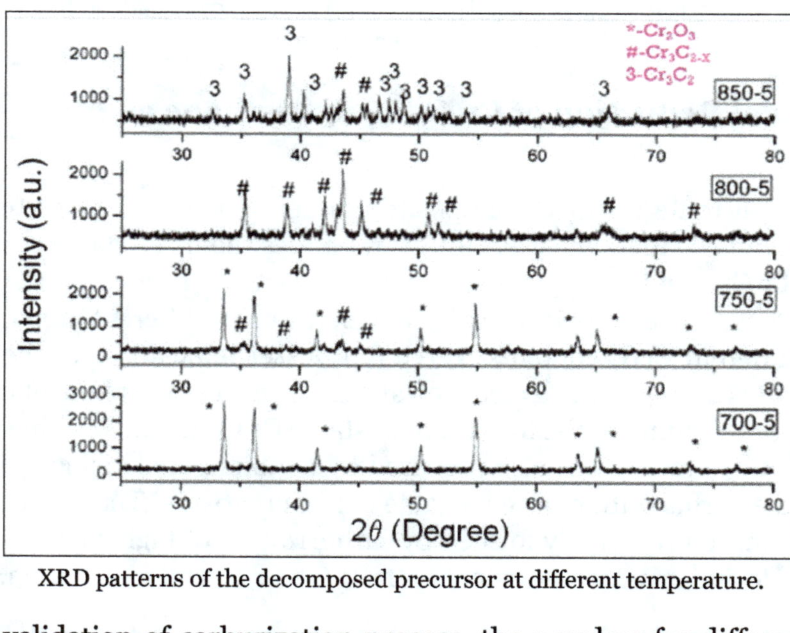

XRD patterns of the decomposed precursor at different temperature.

For further validation of carburization process, the powders for different carburization temperature were characterized by TEM. Figure shows the TEM micrograph of the powder treated at 700 °C, where the particle size is found to be around 40 nm. The diffraction pattern verifies its phase for the chromium oxide Cr_2O_3, (rhombohedral structure). The high resolution image shows a clear atomic image inside the powder. A thin amorphous layer is visible in a different contrast outside the boundary surface. This can be attributed to the adsorption of carbon deposit on the powder surface. Figure shows the TEM and HRTEM of powdered samples carburized at 800 °C for 5 h. The spherical shape has transformed to faceted irregular grain. The diffraction pattern confirms that the phase is attributed to metastable Cr_3C_{2-x}, which is in accordance with the XRD result. The HRTEM image shows that there are several carbon nanofilms formed around the surface of the powder surface. The Fast Fourier transform (FFT) pattern of these layers, shows two obvious halo spots that are attributed to the graphene structure of carbon nanofilms. The TEM micrographs of the powders, carburizated at 850 °C, show that the powder is encapsulated by layers of graphene, as shown in figure. The diffraction pattern verifies that the powder comprises the stable Cr_3C_2 phase.

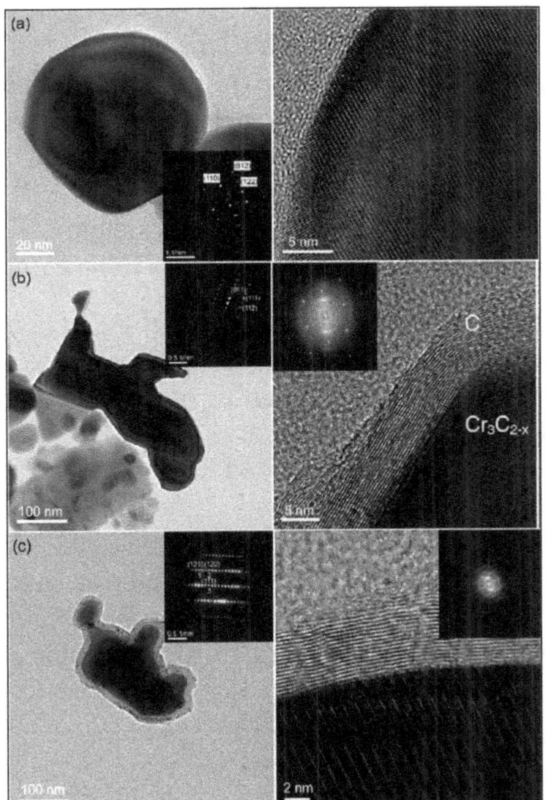

TEM image of decomposed powder after the carbonized treatment
(a) 700 °C/5 h; (b) 800 °C/5 h; (c) 850 °C/5 h.

The phase transformation from Cr_2O_3 to Cr_3C_2 via an intermediate state Cr_3C_{2-x} has been identified using electron-energy loss spectroscopy (EELS). The EELS of prepared samples were characterized by field emission gun scanning transmission electron microscopy (FEG-STEM) equipped with energy dispersive X-ray detector (EDS) and energy filter (Gatan). Figure shows the EELS spectra of Cr L edge for the sample of the decomposed precursor and the samples carbonized at 800 °C and 850 °C. The chromium L edges have features comprising two sharp L_2 and L_3, known as "white lines" and a continuum background following the edge. The sharp double peaks are due to transitions from the $2p^{1/2}$ and $2p^{3/2}$ core levels towards 3d states above E_f. The sharp peaks of L_2

and L_3 at threshold are due to the transitions from the $2p^{1/2}$ core levels to $3d^{3/2}$ states and the $2p^{3/2}$ core levels to $3d^{3/2}$ $3d^{5/2}$, respectively. The L_3/L_2 white line ratio is correlated to the electron occupancy and spin pairing in 3d band. The variations of the ratio are due to the interaction between chromium and surrounding atoms. A change of the L_2, L_3 edge intensities in the samples indicates a variation of the d holes because the edge intensity is proportional to the number of the empty final states available. The ratio of L_3/L_2 of sample shown in figure is higher than that of carbonized samples. According to the results of Arévalo-López & Alario-Franco, the ratio of L_3/L_2 for the Cr_2O_3 is about 1.6 and the ratio of L_3/L_2 for the Chromium carbide is about 1. A comparison between figure shows L_3 edge of the sample carbonized at 850 °C is 577.72eV, while that of sample carbonized at 800 °C is 575.42eV. The shift in energy (2.3 eV) is called the chemical shift. This result is due to the transformation from the metastable carbide to stable carbide, because the EELS represent the difference in energy between a core-level initial state and the lowest energy final state of an excited electron. But the energy loss close to 577.72 eV for the decomposed precursor, which consists of Cr_2O_3 is same as the sample carbonized at 850 °C. It is observed that there is no chemical shift between this two species because Cr_2O_3 and Cr_3C_2 are both chemically stable compounds.

The π electron formed sp^2 bonding observed from the EELS spectra of C K edges as shown in figure. The peak of sp^2 bonding is sharper when the carbonized temperature increases from 800 °C to 850 °C. The sharpening of sp^2 peak is attributed to the formation of graphite like carbon according to Fan et al. In Cr_3C_2, C is graphite like rather than diamond like, which comprising of sp^3 bonded carbon atoms. For the as decomposed precursor, the content of carbon was little and formed amorphous phase. More carbon produced from the pyrolysis of methane coated on the Cr_2O_3 surface and then formed graphite phase and metastable carbide of Cr_3C_{2-x}, when the sample was treated at 800 °C for 5 hours. The increase of treated temperature to 850 °C, enhanced the process of carbonizing reaction leading to more graphite formation followed by transformation of metastable Cr_3C_{2-x} to stable carbide of Cr_3C_2 .

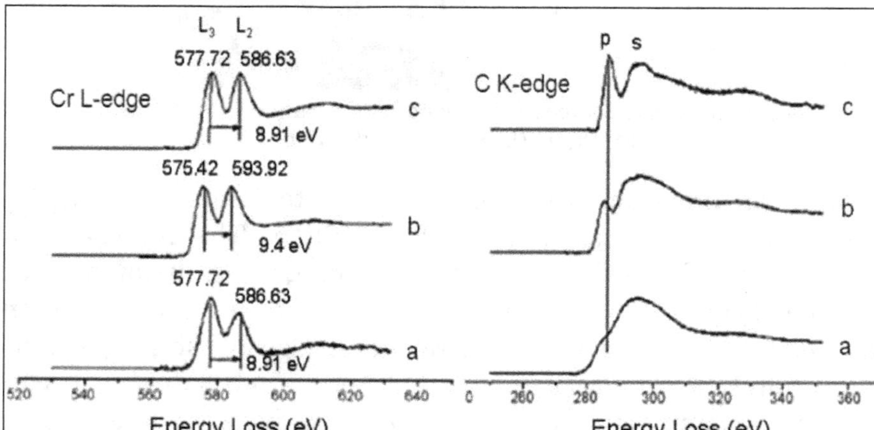

EELS of Cr L edges & C K edges: (a) decomposed precursor, (b) carbonized at 800 °C/5 h (Cr_3C_{2-x}), and (c) carbonized at 850 °C/5 h (Cr_3C_2).

One could hypothesize the following phase transformation steps: In the first step, the methane decomposes to elemental carbon and hydrogen. The carbon reacts with the Cr_2O_3 at temperatures higher than 800 °C, and produce partial Cr_3C_2 phase and $CO_{(g)}$. The $CO_{(g)}$ can also carburize the Cr_2O_3 further. There are three kinds of carbon sources in the system viz. free carbon from the decomposition of methane, CO from above reaction and C-C bonding from the precursor. Finally, the unreacted redundant carbon sources would deposit on the surface of the powder as the lamellar graphene. The graphene deposited during the carburization has an important role for the formation of meta-stable Cr_3C_{2-x} and stable Cr_3C_2. Initially, carbon gets absorb on the surface of chromium oxide at low temperature. As the temperature rises, substantial amount of carbon is available, leading to metastable chromium carbide. Finally, as the temperature increases further, the redundant carbon forms the graphene layers. Further thickening of the graphene films, generates a mechanical stress which assists the transformation of meta-stable Cr_3C_{2-x} state to stable Cr_3C_2 phase.

Microstructure and Mechanical Properties of Cr_3C_2/Al_2O_3 Nanocomposites

The carbonization treatment of initial powders without using methane as the carbon source leads to the formation of (Cr_3C_2, Cr_7C_3)/Al_2O_3 nanocomposites, where as carbonization using methane as carbon source leads to Cr_3C_2/Al_2O_3 nanocomposites, which is the desired material. In order to improve the mechanical properties of these Cr_3C_2/Al_2O_3 nanocomposites, the carbonized powders were undergone hot-press sintering at 1400 °C for 1 hr in vacuum.

Figure shows the SEM micrographs of monolithic Al_2O_3 and Cr_3C_2/Al_2O_3 nanocomposites carbonized at different temperatures for 5 hours and hot-pressed at 1400 °C for 1 hr in vacuum. One can see that there is abnormal grain growth of Al_2O_3 shown in figure. In case of Cr_3C_2/Al_2O_3 nanocomposite, the dispersion secondary phase, Cr_3C_2 is uniform and the abnormal grain growth has been inhibited. The grain size distribution of monolithic Al_2O_3 and Cr_3C_2/Al_2O_3 nanocomposites has been shown in figure. The nanocomposite exhibits low grain size than that of pure alumina. It implies that addition of Cr_3C_2 decreases the grain size and inhibits the growth of alumina matrix.

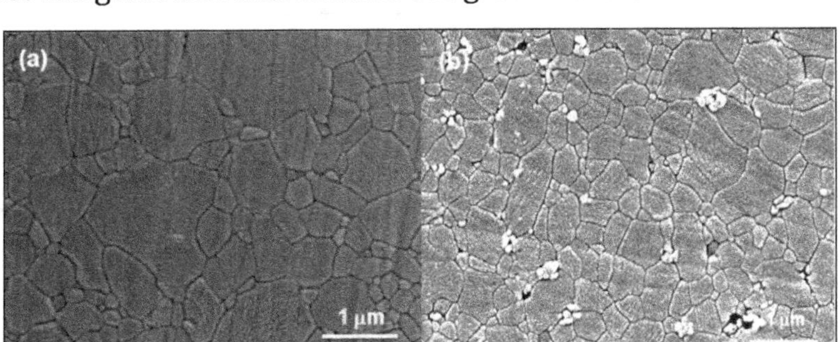

SEM micrographs of (a) monolithic Al_2O_3 and (b) Cr_3C_2/Al_2O_3 nanocomposites by carbonized at different temperatures for 5 hours and hot-pressed at 1400 °C for 1 hr in vacuum.

The TEM micrograph of Cr_3C_2/Al_2O_3 nanocomposites shown in figure illustrates two types of inclusion of Cr_3C_2 onto the Al_2O_3 matrix i.e. inter-type inclusion and intra-type inclusion. Figure shows the SEM micrographs of the fracture surface of monolithic Al_2O_3 and Cr_3C_2/Al_2O_3 nanocomposites under the same carbonization condition. The facture surface of pure alumina is intergranular in nature, where as transgranular facture surface is observed in Cr_3C_2/Al_2O_3 nanocomposites.

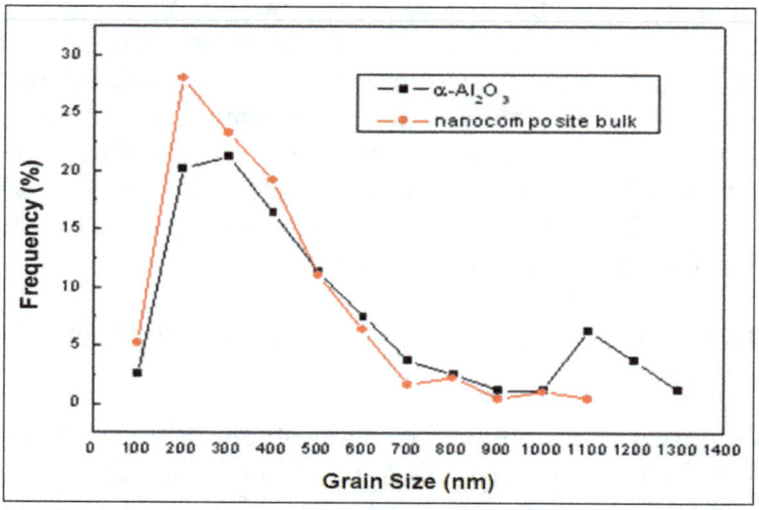

Grain size distribution of (a) monolithic Al_2O_3 and (b) Cr_3C_2/Al_2O_3 nanocomposites prepared by carbonization at different temperatures for 5 hours followed by hot-press sintering at 1400 °C for 1 hr in vacuum.

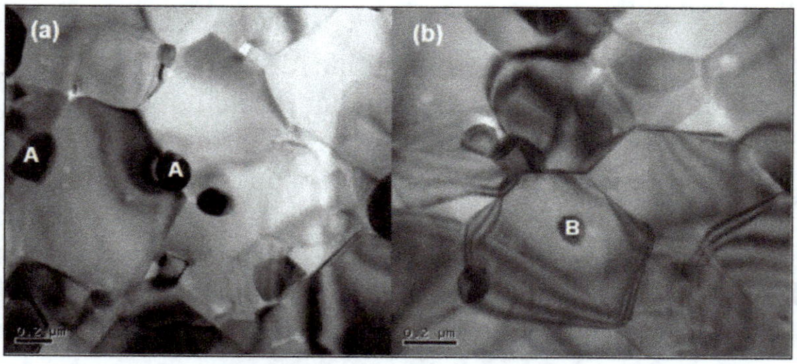

TEM micrographs of Cr_3C_2/Al_2O_3 nanocomposites (a) inter-type inclusion in A (b) intra-type inclusion in B.

The basic mechanical properties such as hardness, fracture strength and toughness of pure alumina as well as of this nanocomposite have been analyzed and mentioned in table. The fracture strength was determined by 4-point bending (Shimadzu AG-IS 100KN, Japan). The toughness was measured by Single-Edge-Precracked Beam (SEPB). A precrack with depth of 0.254mm was conducted in the center of sample by a diamond cutter with a thickness of 0.15mm. It is observed that the nanocomposite exhibits improved hardness, fracture strength and facture toughness in comparison to monolithic alumina.

The relationship between the microstructure and improved mechanical property has been investigated and described below.

SEM micrographs of the fracture surface (a) monolithic Al_2O_3 and (b) Cr_3C_2/Al_2O_3 nanocomposites prepared by carbonized at different temperatures for 5 hours hot-pressed at 1400 °C for 1 hr in vacuum.

Table: Mechanical property measurements on monolithic Al_2O_3 and Cr_3C_2/Al_2O_3 nanocomposites.

Specimen	Hardness	Strength (4 pt-bending)	Toughness (Indentation Fracture method)
Al_2O_3	18.4 GPa	375 MPa	4.0 $MPa.m^{1/2}$
Cr_3C_2/Al_2O_3	19.8 GPa	520 MPa	5.5 $MPa.m^{1/2}$

The nanocomposites consisting of Cr_3C_2 particles dispersed within Al_2O_3 matrix grains, result in the generation of thermally induced residual stresses after hot-pressed sintering at 1400C. The mismatch in thermal expansion co-efficient between the matrix (α-Al_2O_3= $8.4*10^{-6}$/ °C), and the dispersed particles (Cr_3C_2=$11.2*10^{-6}$/ °C) yields highly localized residual stresses around the particles. These stresses reduce quickly as distance from the boundary increases because of the nano-sized particles, which can generate only small defects such as dislocations shown in figure in close vicinity to the particles. In the monolithic alumina, crack propagates along with the grain boundaries due to the existence of defects and the compress residual stresses resulting from anisotropic thermal expansion, Young's modulus along the crystal axes, and crystallographic misorientation across the grain boundaries. The fracture toughness of grain boundaries is usually lower than that within the grains. However, in nanocomposites, the dislocations around the Cr_3C_2 particles release residual stresses in the Al_2O_3 matrix. Consequently, the defect size along the grain boundaries is reduced. Also, the dislocations serve as origins of small stress concentrations, and create nano-cracks around the propagating crack tip. These nano-cracks slightly reduce the strength of the alumina matrix, while reduction of both the residual stresses along the grain boundaries and the strength in the matrix lead to a change in the fracture mode from that of the intergranular fracture in monolithic alumina to that of transgranular fracture in nanocomposites, as shown in figure. In addition, due to the effect of nano-cracking, the step-wise fracture surface rather than a planar cleavage plane is observed in the nanocomposites, as shown in figure. The reinforcement of the Cr_3C_2 particles, small grain size of Al_2O_3 matrix and reduction of both the defect size along the grain boundaries and the tensile residual stresses in the matrix grains by dislocations result in improvement of the strength of nanocomposites.

In the case of Cr_3C_2-Al_2O_3 solid solution described in the previous section, the Cr^{3+} ions increase the growth rate of Al_2O_3 grain because of the coherency strain energy at grain boundary. However, this phenomenon is not found in the nanocomposites. It is attributed that the secondary particles of Cr_3C_2 inhibits the growth of alumina matrix. The fracture strength is increased by the grain-boundary modification caused the compressive stress generated by the substitution of bigger Cr^{3+} ion (0.076 nm) in place of smaller Al^{3+} ion (0.068nm). The localized compressive stress helps in strengthening the grain boundary, and then the fracture strength increases by this effect. The higher toughness of the nanocomposites relative to monolithic alumina may be connected with the stepped transgranular cleavage. The direction of crack propagation within the matrix grains is affected by these dislocations.

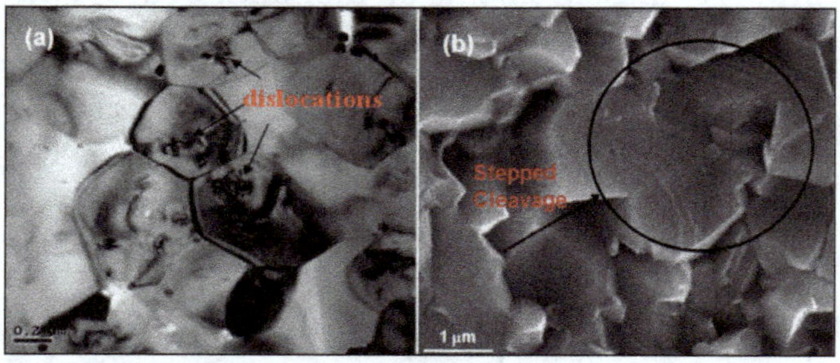

a) dislocations and (b) stepped cleavage observed in Cr_3C_2/Al_2O_3 nanocomposites.

Densification Behavior of Al_2O_3-Cr_2O_3/Cr_3C_2 Nanocomposites Prepared by Spark Plasma Sintering

Compared with conventional hot pressed sintering, spark plasma sintering (SPS) has recently attracted increasing interest due to its ability of rapidly heating the powder compact to high temperatures and consolidating it to high densities within very short time. SPS utilizes applied pressure along with temperature and a pulsed direct current to densify materials. SPS is heralded for its promise in retaining nanostructure features, producing grain boundaries devoid of impurities as well as many other benefits.

As described earlier, the initial powders used for this sintering process were Al_2O_3-Cr_2O_3, which were prepared by metal organic chemical vapor deposition (MOCVD) in a fluidized bed. The nanosized composite powders fabricated in the fluidized bed for 30 and 60 minutes were named as S-30 and S-60, respectively. The pressure of reaction chamber was controlled at 10 torr, and the reaction temperature was kept at 300 °C. The fluidized powders were densified by a SPS process (SPS-515S, Shumitomo, Japan). The 1.5gram of powder (pure Al_2O_3, S-30, S-60) was put into a graphite mold of 15.5mm diameter. The uniaxial pressure of 50Mpa was imposed on the powder and the vacuum level was less than 6 Pa. The heating rate from room temperature to 600 °C and from 600 °C to sintering temperature was 200 °C/min and 100 °C/min, respectively with a holding time of 10 minutes. There were four different sintering temperature conditions of 1200 °C, 1250 °C, 1300 °C and 1350 °C.

Effect of SPS Sintering Temperature on the Color Changes of Compacts

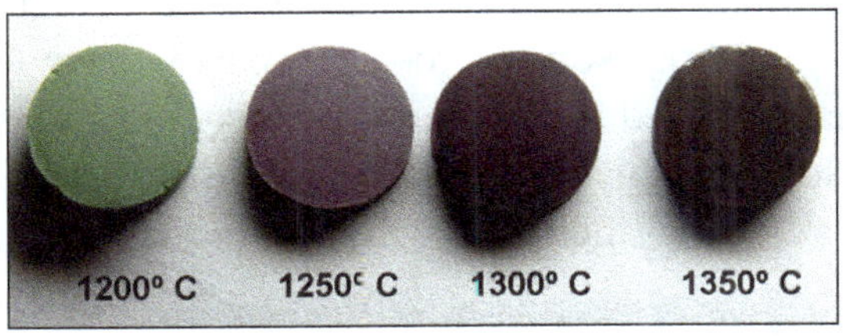

Optical micrograph of compacts prepared by different SPS temperatures from 1200 °C to 1350 °C.

The color of compacts prepared under different SPS conditions is shown in figure. It is observed that the color of compacts sintered at 1200 °C is green; the color of 1250 °C compact is purple; while that of 1300 °C and 1350 °C are black. Figure indicates the XRD patterns of the compacts processed by various SPS temperatures. Besides Al_2O_3 peaks, the Cr_2O_3 peaks are found in the compact sintered at 1200 °C. Compared with the peak of pure-Al_2O_3, the Al_2O_3 peaks of compact sintered at higher temperature have slight shifted to lower angles. It means only a little solid-solution of Al_2O_3-Cr_2O_3 has formed in this compact. This result is special to the compacts fabricated by conventional densification methods in which significant pink solid-solution of Al_2O_3-Cr_2O_3 was formed. According to Bondioli et al. the complete ranges of Al_2O_3-Cr_2O_3 solid solution can be obtained for the reaction temperature higher than 1000 °C.

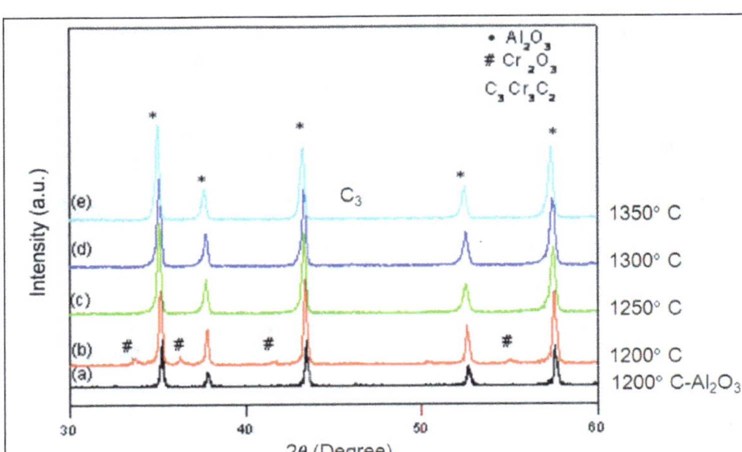

XRD patterns of (a) pure Al_2O_3 sintered at 1200 °C, and the compact at (b) 1200 °C, (c) 1250 °C, (d) 1300 °C and (e) 1350 °C for 10 min by SPS.

It is suggested that due to the higher heating rate and short holding time, the most of Cr^{3+} doesn't have enough time to diffuse into Al_2O_3 matrix and remains as Cr_2O_3 phase after the SPS process. The color of pure Cr_2O_3 is green, and so the composite is with green appearance. For the compact sintered at 1250 °C, the Cr_2O_3 peaks are disappeared in the XRD pattern because more Cr_2O_3 react with Al_2O_3 matrix to form pinker solid solution in this higher sintering temperature. According to Equation 4 given in, chromium carbide would form probably when the temperature is higher than 1150 °C. It is believed that there is little black carbide formed when sintered at 1250 °C, so the compact exhibits purple color. It is reasonable that the more Cr^{3+} diffuse into Al_2O_3 matrix with the increase in sintering temperature. In comparison to other compacts, more Al_2O_3 peaks of the compact sintered 1350 °C have shifted to lower angles observed from XRD patterns. It indicates that more solid solution is formed at a higher temperature. The amount of the solid solution of Al_2O_3-Cr_2O_3 and Cr_3C_2 increases with the increase of sintering temperature.

$$3CrO_{3(s)} + 13C_{(a)} \rightarrow 2Cr_3C_{2(s)} + 9CO_{(g)}$$
$$\text{Where}, \Delta G = 261313.95 - 188.5T(\text{cal})$$

The black color of the compact prepared at 1350 °C is relevant to the formation of chromium carbide. More Cr_3C_2 is formed as the temperature is over than 1150 °C. Cr_3C_2 nanoparticle is observed for the compacts prepared at 1300 °C and 1350 °C. The TEM diffraction pattern shows the particle with Cr_3C_2 phase as shown in figure. Although the peak of CrC_2 peaks are not obvious in the XRD patterns in figure, but based on the $Cr_2O_3(s)$ + 13/3C(s) → 2/3$Cr_3C_2(s)$ + 3CO(g), sintering at a higher temperature is beneficial to form Cr_3C_2. Cr_3C_2 peaks are easy to be found in the XRD patterns as compacts prepared by HP densification in which the vacuum maintained at about 10^{-5} torr, however in this SPS process that is about 6 pa (~10^{-2} torr). According to the report of Chu & Rahmel, carbon would be consumed by oxygen in a higher oxygen pressure in SPS process and inhibits the formation of Cr_3C_2.

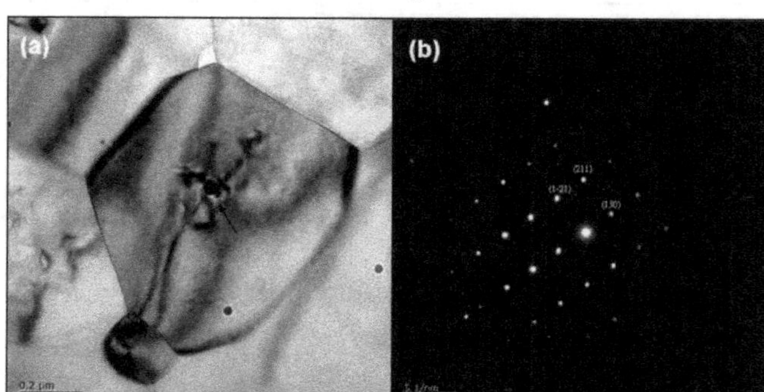

TEM micrographs of nanocomposites sintered at 1350 °C
(a) bright field image; (b) SADP of particle in (a).

Effect of SPS Sintering Temperature on Shrinkage Behavior

Figure shows the sintering shrinkage curves of the pure Al_2O_3, nanocomposite powders of S-30 and S-60 during the SPS densification in the temperature range 1200 °C ~1350 °C with holding time of 10minutes. The y-axis $\Delta L/L_0$ (%) represents the shrinkage profile directly during densification of the powders in real-time. The trend of Al_2O_3 curve becomes flat at 1200 °C, but the curves of S-30 and S-60 still trended up. This means the shrinkage Al_2O_3 has already stopped, but S-30 and S-60 continue shrinking during the period of holding time.

This result is consistent with the apparent density shown in figure. It shows that pure Al_2O_3 exhibits a higher density than S-30 and S-60, for the sintering temperature at 1200 °C and 1250 °C. The amount of shrinkage of S-30 is more significant than S-60. The temperature at which the shrinkage vanishes for S-30 and S-60 is 1300 °C and 1350 °C, respectively as shown in figure. The curves of pure Al_2O_3, S-30 and S-60 are trended down for the holding temperature at 1350 °C. This signifies that all the powder compacts undergone thermal expansion after completion of the sintering shrinkage. The densities of S-30 and S-60 are increased significantly when the temperature rises to 1300 °C and 1350 °C.

Figure illustrates that the pure Al_2O_3 starts to densify at about 950 °C and finishes at about 1300 °C. For S-60 and S-30, the densification temperature starts at about 1050 °C and 1100 °C, separately, and finishes at 1350 °C for both. Actually S-60 finishes its densification after holding about 50 seconds. The second phase particles decrease the diffusivity of grain boundary and lattice, and result in the retardation of densification. According to XRD spectrum shown in figure, the delay on the densification process for the nanocomposites is due to the Cr_2O_3 particles coated on Al_2O_3 as sintered at 1200 °C. Cr_2O_3 plays as a second phase to inhibit the grain growth during densification. While some of Cr_2O_3 changes to Cr_3C_2 to be the second phase as temperature increases to 1300 °C and 1350 °C.

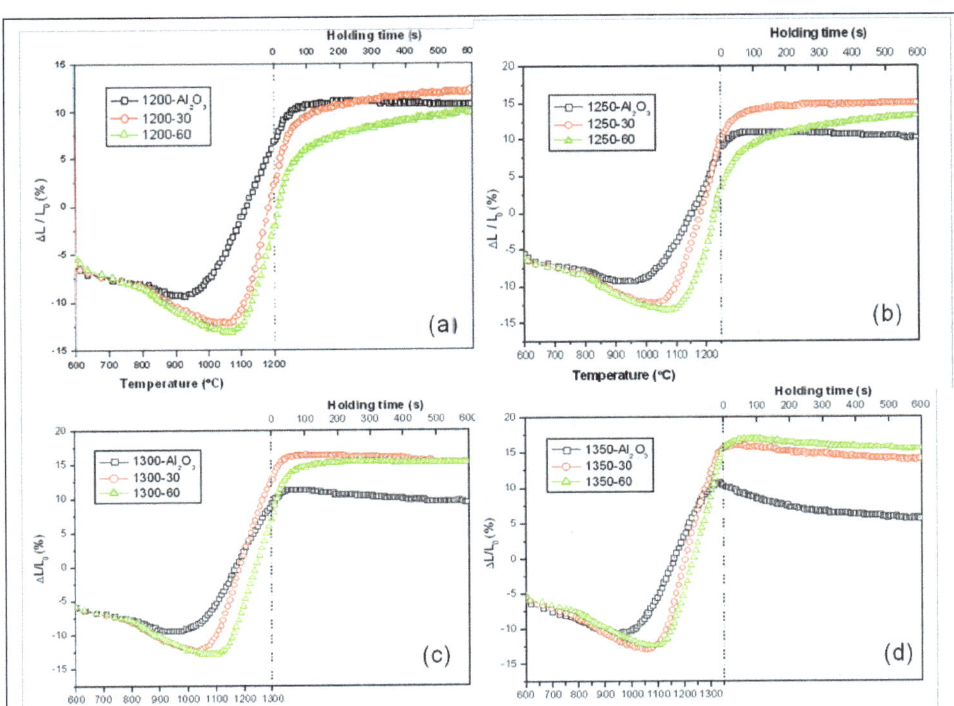

Sintering shrinkage curves of Al_2O_3, S-30 and S-60 sintered at (a) 1200 °C, (b) 1250 °C (c) 1300 °C (d) 1350 °C for 10min by SPS.

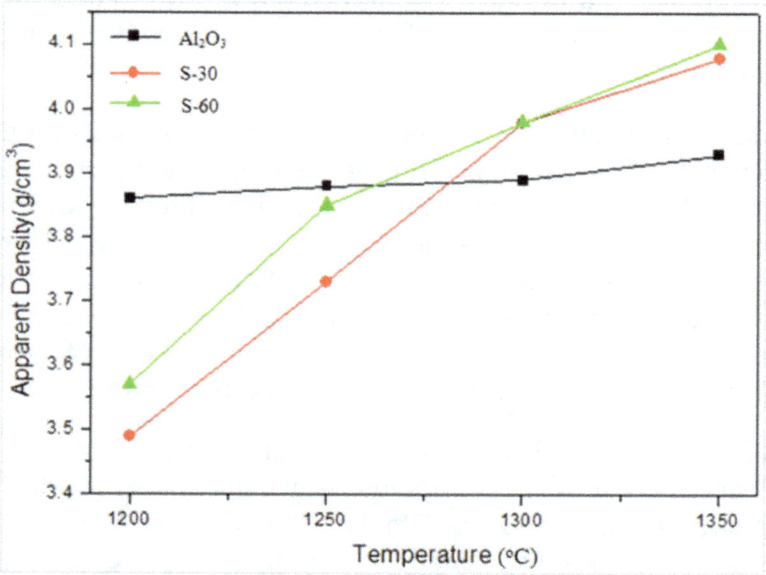

The apparent density of the specimen of Al_2O_3, S-30, and S-60 prepared at a different SPS sintering temperature from 1200 °C to 1350 °C.

Effect of SPS Sintering Temperature on Microstructures and Mechanical Properties

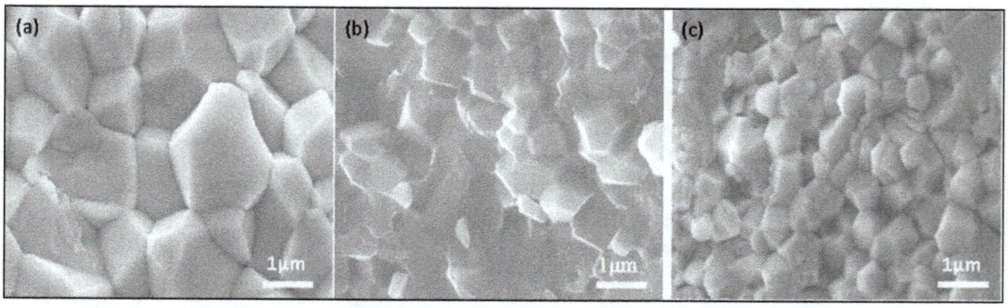

SEM micrographs of fracture surface of 1350 °C fully dense SPS specimen of (a) Al_2O_3, (b) S-30, (c) S-60.

The fine grains of Al_2O_3-Cr_2O_3/Cr-carbide nanocomposites were prepared by employing spark plasma sintering (SPS) technique. The basic mechanical properties like hardness, fracture strength and toughness were analyzed. The fracture surfaces of the nanocomosites are shown in figure. It indicates that the pure alumina has larger grain size than that of S-30 and S-60 densification samples. The nanosized reinforced particles (Cr_3C_2) are found in the alumina matrix. S-60 densification sample has smallest grain size due the more amounts of secondary phase particles to inhibit its grain growth.

The hardness, fracture strength and toughness of the S-30 and S-60 nanocomposites are present in figure. The S-60 nanocomposite exhibits highest fracture strength and toughness. The increase in hardness, fracture strength and toughness of nanocomposites over monolithic alumina is due to the strengthening of secondary phase of Cr_3C_2 particles and solid solution of Al_2O_3-Cr_2O_3. The co-relation between the microstructure and mechanical properties is similar to that of nacocomposites prepared using hot press sintering.

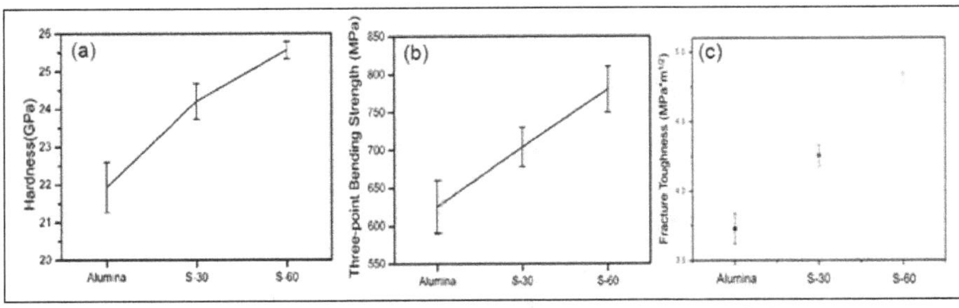

a) Hardness, (b) Fracture Strength and (c) Toughness of Chromium Carbide/Alumina nanocomposites sintered at 1350 °C for 10min by SPS.

Nanoporous Materials

4

- Microporous Material
- Mesoporous Material
- Macroporous Materials
- Example of Nanoporous Solids

The organic or inorganic materials comprising of pores of size less than 100nm that support the porous structures are termed as nanoporous materials. They can be branched into three categories – microporous materials, mesoporous materials and macroporous materials. The topics elaborated in this chapter will help in gaining a better perspective of nanoporous materials.

Nanoporous materials consist of a regular organic or inorganic framework supporting a regular, porous structure. The size of the pores is generally 100 nm or smaller. Most nanoporous materials can be classified as bulk materials or membranes.

Classically, porous matter is seen as a material that has voids through and through. The voids show a translational repetition in 3-D space, while no regularity is necessary for a material to be termed "porous." A typical and relatively simple porous system is one type of dispersion classically described in colloid science, namely a foam or, better, a solid foam. In correlation with this, the most common way to think about a porous material is as a material with gas-solid interfaces as the most dominant characteristic. This already indicates that classical colloid and interface science as the creation of interfaces due to nucleation phenomena (in this case nucleation of wholes),decreasing interface energy, and stabilization of interfaces is of elemental importance in the formation process of nanoporous materials. These factors are often omitted because the nal products are stable. Indeed, they are just metastable. This metastability is due to the rigid character of the void-surrounding network, which is covalently cross-linked in most cases. However, it should be noticed that most of the porous materials reported in the literature or those that are of high technical relevance are not stable by thermodynamic means. As soon as kinetic energy boundaries are overcome, materials start to break down. One example should elucidate this. Porous silica, for instance, is just metastable. As soon as the temperature is raised and the melting point is reached, primary particles in the network begin to fuse and it comes to phase separation into a nonporous silica phase. Finally, at very high temperatures, the thermodynamic stable phase of SiO_2 quartz emerges.

Control over interface energy and metastabilization of nanodimensional holes becomes of special importance when the task is to produce nanoporous materials. For this review, the whole range of pores from 1 nm to 1000 nm is interpreted as nanoporous materials.

Accordingly, porous materials might be classif ed by the size of pores or may be distinguished by different network materials.

Relevance of Nanoporous Materials

One can find a whole variety of nanoporous materials in nature executing many different functions. The most common task for nanoporous materials in nature is to make inorganic material much lighter while preserving or improving the high structural stability of these compounds. Often, by filling the voids between inorganic matter the desired properties of the hybrid materials exceed the performances of the pure compounds by several orders of magnitude. Nanoporous materials in nature are organicinorganic hybrids. Naturally occurring materials exhibit synergistic properties. Neither the organic material filling the void nor the inorganic network materials are able to achieve comparable performances by themselves. Examples of these materials are found for diatoms, radiolarii, or the famous abalone shell. Even in the human body, life without bones would not be conceivable.

It is seen that complex mechanisms are involved in the formation of these hierachical materials. Similar to the structure motives on different length-scales cells, vesicles, supramolecular structures, and biomolecules are involved in the structuring process of inorganic matter occurring in nature. This process is commonly known as biomineralization. It is often not seen in this relation, but it will be shown later that ordered porous materials, and therefore artificial materials, are constructed according to very similar principles. A completely different area where nanoporous materials are highly important is in the lungs, where a foam with a high surface area permits sufficient transfer of oxygen to the blood. Even the most recent developments in nanoporous materials, such as their application as photonic materials,are already present in nature; the color of butterfly wings, for instance, originates from photonic effects.

It can be concluded that nature applies the concept of nanoporous materials (either filled or un filled) as a powerful tool for constructing all kinds of materials with advanced properties. It is therefore not surprising how much research has recently been devoted to porous materials in different areas such as chemistry, physics, and engineering. The current interests in nanoporous materials are now far behind their size-sieving properties.

Classification by Pore Size

The pore-size regime for nanoporous materials ranges from the 1-nm region to 1000 nm (+ 1 μm). According to IUPAC, three distinctions can be made:

- Microporous materials 0–2-nm pores.
- Mesoporous materials 2–50-nm pores.
- Macroporous materials >50-nm pores.

A comparison of these different pore systems is given in figure. However, in these cases ordered pore systems are shown. Order is no premise at all. Microporous, mesoporous, or macroporous materials might have random systems of pores. In the field of mesoporous materials in particular, the mistake is often made of mixing up mesoporous and ordered materials

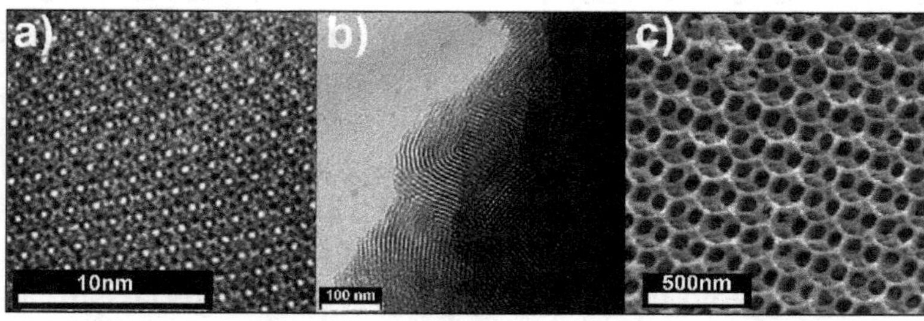

Three different types of nanoporous silica materials. (a) A microporous material: zeolite Y. (b) A mesoporous materials. (c) A macroporous material.

Ordered porous materials are judged to be much more interesting because of the control over pore sizes and pore shapes. Their disordered counterparts exhibit high polydispersity in pore sizes, and the shapes of the pores are irregular (a nice example of such a material can be found in the literature). Ordered porous materials seem to be much more homogeneous. But one has to be careful with this statement because it is often led by aesthetic arguments caused by the beauty of images. In many cases a material possesses more than one porosity. This could be:

- For microporous materials: An additional meso- or macroporosity caused by random grain packing.
- For mesoporous materials: An additional macroporosity caused by random grain packing, or an additional microporosity in the continuous network.
- For macroporous materials: An additional meso- and microporosity.

These factors should be taken into consideration when materials are classified according to their homogeneity. A material possessing just one type of pore, even when the pores are disordered, might be more homogenous than one having just a fraction of nicely ordered pores.

A good tool for deciding and classifying porous materials according to the size of their pores is sorption analysis. Indeed, pores of different sizes lead to totally different characteristics in sorption isotherms.

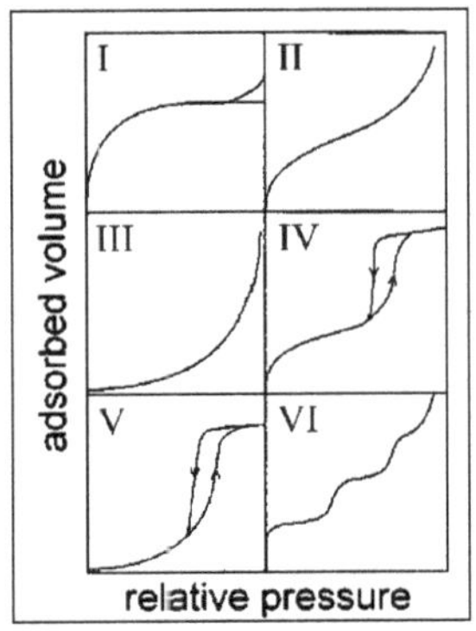

Sorption isotherms of microporous materials (I), mesoporous materials (V), and macroporous materials (III).

The correlation between the vapor pressure and the pore size is given by the Kelvin equation (shown here in a slightly modified form):

$$r_p\left(\frac{p}{p_o}\right)=\frac{2\gamma V_L}{RT\ln(p/p_o)}+t\left(\frac{p}{p_o}\right)$$

where $r_p \triangleq$ pore radius, $\gamma \triangleq$ surface tension, $t \triangleq$ thickness of the adsorbate film, $V_L \triangleq$ molecular volume of the condensate.

Therefore, the isotherms of microporous materials show a steep increase at very low pressures (relative pressures near zero) and reach a plateau quickly. Mesoporous materials are characterized by a so-called capillary condensation step and a hysteresis (a difference between adsorption and desorption). Macroporous materials show a single or multiple adsorption steps near the pressure of the standard bulk condensated state (relative pressure approaches one).

Classification by Network Material

One of the most important goals in the field of nanoporous materials is to achieve any possible chemical composition in the network materials "hosting" the pores. It makes sense to divide the materials into two categories: (a) Inorganic materials and (b) organic materials. Among the inorganic materials, which is the larger group, we find:

- Inorganic oxide-type materials: This is the field of the most commonly known porous silica, porous titania, and porous zirconia materials.
- A category of its own is given for nanoporous carbon materials: In this category are the highly important active carbons and some examples for ordered mesoporous carbon materials.

- Other binary compounds such as sulfides, nitrides etc. Into this category also fall the famous $AlPO_4$ materials.
- There are already some examples in addition to carbon where just one element (for instance, a metal) could be prepared in a nanoporous state. The most prominent member of this class of materials is likely to be nanoporous silicon, with its luminescent properties.

Modern Analysis of Nanoporous Materials

Gas Adsorption Methods

Gas sorption represents a widely used technique for characterizing microand mesoporous materials and provides porosity parameters such as pore size distributions, surface areas, and pore volumes.

In a typical sorption experiment the uptake of gases such as nitrogen, krypton, and CO_2 is measured as a function of relative pressures $p/p_0 < 1$ at constant temperature. p and p_0 are the equilibrium vapor pressures of the liquid in the pores and that of the bulk liquid, respectively. The interaction between the pore walls and the adsorbate is based on physisorption (van der Waals interaction) and leads to the formation of adsorbate layers at low $p/p_0 < 1$. Simplistically, the macroscopic laws of classical thermodynamics predict that the confinement of pores with radii on the nanometer scale leads to the condensation of gas inside the pores at a pressure smaller than p_0. In a typical sorption experiment the adsorbed volume is plotted versus $p/p_0 < 1$, and this "sorption isotherm" is the superposition of different uptake mechanisms. At low p/p_0 adsorption in micropores takes place, which is supposed to be a process of volume filling rather than capillary condensation. The shape of an isotherm itself distinguishes between representative types of nanoporous materials, based on the classifications by IUPAC or de Boer. The isotherms of microporous materials are characterized by a steep increase of the isotherm at low $p/p_0 < 1$, ending up in a plateau at larger $p/p_0 < 1$. In a mesoporous substrate, with increasing values of $p/p_0 < 1$, a liquid-like adsorbate lm of statistical thickness $t(p/p_0)$ is formed on the pore walls. At a certain pressure, capillary condensation takes place, filling the mesopores with liquid, which is apparent in isotherms as a pronounced increase in the adsorbed amount. The total pore volume ("porosity") is given by the overall uptake of adsorbate. Zsigmondy proposed the first explanation of the capillary condensation in a single in nite cylindrical mesopore based on the macroscopic Kelvin equation $(p/p_0) = 2\sigma V_L / RTr_m$, where V_L and ' are the molal liquid volume and the liquid-gas surface tension at temperature T, respectively, and r_m is the mean radius of curvature of the liquid/gas interface (cylindrical for the condensation, spherical for desorption).Based on the classical treatment of Cohan and the Kelvin equation, the condensation of a liquid in a nanoporous material at a certain p/p_0 can be related to the corresponding mesopore size, thus also providing a pore size distribution (PSD). Since the condensation starts at a relative pressure p/po , where the walls are covered by a film of thickness $t(p/p_0)$, in the so-called modified Kelvin equation $2/r_m$ is replaced by $f/(r-t(p/p_0))$, where r is the "true" mesopore radius and f is the meniscus shape factor, which is 1 or 2 for the filling or emptying of the mesopore, respectively. This procedure represents the basis of the well-known "Barrett-Joyner-Halenda (BJH)" method, currently the procedure most frequently used to determine PSDs. The dependence $t(p/p_0)$ can be described by the approach of Frenkel-Halsey-Hill describing the sorption on nonporous silica. An improved treatment, originating with Derjaguin,40 takes into account the influence of surface forces on adsorbed film equilibrium and stability, which leads to predictions for capillary condensation and desorption pressures that are substantially different from those of Cohan's theory. In addition, the pore geometry significantly affects thermodynamic properties of confined fluids and their adsorption behavior. The confinement is stronger in spherical pores compared with cylindrical pores of the same diameter, leading to a shift of the capillary condensation to lower p/p_0; therefore the aformentioned procedures are restricted to cylindrical pores. Broekhoff and de Boer described the condensation in spherical ink-bottle pores, connected by narrow cylindrical windows, based on the Kelvin-Cohan approach, also predicting the appearance of hysteresis.

The progress in synthesizing mesoporous materials with well-de ned pore morphologies in terms of uniform mesopore sizes and pore shapes allowed the testing, optimization, and further development these classical approaches. Based on the BJH method, Kruk and Jaroniec determined the mesopore size of MCM-41 by using the "KJS" approach, which corrects the Kelvin equation by an empirical additive constant (0.3) for the mesopore radius, which, however, does not have a theoretical foundation: $r(p/p_0)[nm] = 2\sigma V_L / RT \ln(p/p_0) + t(p/p_0) + 0.3$.

Moreover, Kruk and Jaroniec used basic geometrical relationships to determine the mesopore diameter of MCM-41 based on the (100) interplanar spacing as determined from X-ray diffraction and the mesopore volume, assuming an array of uniform pores.

A further approach to calculating pore volumes and mesopores sizes is based on the Gurvich approach, in which the adsorbed volume $V_{0.4}$ at $p/p_0 \approx 0.4$ is used to calculate the pore diameter by $D = 4V_{0.4}/S_{BET}$, where S_{BET} is the specific surface area obtained from the method introduced by Brunauer, Emmet, and Teller (BET).The BET method is based on the assumption that multiplayer formation takes place prior to capillary condensation and that the equilibrium state is characterized by different rate constants for adsorption and desorption for the mono and multilayer. The adsorbed amount n_a, p/p_0, and the monolayer capacity n_m are related by the BET equation, from which the specific surface area S_{BET} is calculated as $S_{BET} = n_m a_m N_A$, where a_m is the area of an adsorbate molecule and N_A is Avogadro's number.

Further widely used methods for determining structural parameters are comparative plots such as the t-plot and α-plot methods. Typically, the amount adsorbed on the porous solid under study is plotted as a function of the amount adsorbed on an ideally nonporous reference solid with similar surface characteristics, providing parameters such as the overall pore volume, specific surface area, and micropore volumes. This procedure has been used to determine microporosity in mesoporous silicas such as SBA-15 showing nonnegative intercepts in the comparative plot. Based on comparative plots, Jaroniec et al. concluded that MCM-41 does not contain extra microporosity.

The procedures described above are not appropriate for the characterization of microporous materials, particularly microporous carbons and zeolites. Among the most frequently used evaluation procedures are the phenomenological models based on Dubinin's theory of volume filling of micropores, such as the Dubinin-Radushkevich (DR), Dubinin-Astakhov, and Dubinin-Stoeckli equations. Another approach is the Saito-Foley method, which is an extension of the Horvath-Kawazoe method.

It has turned out that the macroscopic thermodynamics of the classical methods described above do not provide reliable descriptions of materials with mesopore sizes below about 4 nm for oxidic materials. One of the main shortcomings of these approaches lies in the nonconsideration of fluid-wall interactions. Recent progress in understanding capillary condensation deals with molecular level models. The methods of the grand canonical Monte Carlo (GCMC) simulations, molecular dynamics, and density functional theory (DFT) allow direct modeling of capillary condensation/desorption phase transitions and are capable of generating hysteresis loops of simple fluids sorbed in model pores. Neimark and Ravikovitch have shown that the nonlocal density functional theory (NLDFT), with properly chosen parameters of fluid- fluid and fluid-solid intermolecular interactions, quantitatively predicts desorption branches of hysteretic isotherms of nitrogen and argon on reference MCM-41 samples with pore channels narrower than 5 nm. This method was tested against Monte Carlo simulations and was shown to provide reliable pore sizes and wall thicknesses in MCM-41 materials. Neimark and Ravikovitch extended the NLDFT approach to MCM-41- type materials such as SBA-15 with pore sizes larger than 5 nm, and, in a recent study, they applied the NLDFT approach successfully to cage-like mesopore silicas such as SBA-1, SBA-2, and SBA-12. Moreover, the NLDFT predictions of equilibrium and spontaneous capillary condensation transitions for pores wider than 6 nm were well approximated by the macroscopic equations of the Derjaguin-Broekhoff-de Boer theory, while the results of the traditional Cohan equation (BJH method) were shown to be significantly in error.

In spite of the recent progress in the theoretical understanding of sorption phenomena, certain issues of sorption are still unclear:

- The nature of the hysteresis in mesoporous materials is still subject to intensive theoretical and experimental research and is not yet fully understood. While certain materials with small mesopores such as MCM-41 or MCM-48 exhibit equilibrium capillary condensation, not showing a hysteresis loop at all in nitrogen sorption experiments at T = 77 K, mesoporous silicas like SBA-15 show prounounced hysteresis loops with parallel adsorption and desorption branches. In contrast, other types of mesoporous silicas show a steep decrease in the desorption curve. Recent NLDFT studies were able to shed some more light on the hysteresis phenomenon. It was concluded that in the range of pore sizes greater than 5 nm, for MCM-41-like materials the experimental desorption branch corresponds to equilibrium evaporation, while the capillary condensation branch corresponds to spontaneous (spinodal) condensation. Similar results were obtained for cage-like pores in siliceous materials. Froba and co-workers carried out a systematic study of adsorption/desorption phenomena in MCM-48 by using different gases such as nitrogen, argon, and krypton at various temperatures. It was concluded that hysteresis is a function of the pore size and the temperature.

- The simultaneous presence of mesopores and irregular micropores in the walls, as shown for SBA-15 and other silicas obtained from PEO-containing templates, severely impedes the evaluation of surface areas and the quantification of the micropore volumes. Although it was shown that the use of comparative plots underestimates the microporosity, t-plots and α-plots are still frequently used for these materials.
- The dependence $t(p/p_0)$ is still a matter of discussion because of the lack of independent techniques for determining lm thicknesses, and usually reference data from nonporous materials are used. An exact knowledge of $t(p/p_0)$ is needed for the determination of mesopore sizes by classical methods and for testing DFT models. Smarsly et al. have determined $t(p/p_0)$ by a combination of nitrogen sorption and small-angle neutron scattering.

In mercury porosimetry (MP), gas is evacuated from the sample, which is then immersed in mercury, and an external pressure is applied to gradually force the nonwetting mercury into the sample. By monitoring the incremental volume of mercury intruded for each applied pressure, the pore size distribution of the sample can be estimated in terms of the volume of the pores intruded for a given diameter D. The evaluation of pore sizes from MP is based on the Washburn equation, quantifying the pressure p required to force a nonwetting fluid into a circular cross-sectional capillary of diameter D. MP allows the determination of PSD between 3 nm and 200 nm and is therefore inappropriate for microporous materials but more suitable for pore sizes above ca. 30 nm compared with nitrogen sorption. As further parameters, the total pore volume V_{tot} is accessible from the total intruded volume of mercury at the highest pressure determined, and the total pore surface S is calculated from $S=(1/\gamma|\cos\theta)\int_0^{V_{tot}} p\,dV$. MP has inherent shortcuts in determining mesopore sizes, especially below 20 nm. During the measurement, high pressures used to force mercury into small pores may compress the sample. Damage or compression of highly porous silica has been reported previously. In addition, mercury porosimetry overestimates the volume of the smallest pores in the case of ink-bottle-shaped pores by the small openings.

Electron Microscopy

Electron microscopy (transmission electron microscopy (TEM) for micro- and mesoporous samples and scanning electron microscopy (SEM) for macroporous samples) is an indispensable tool for the investigation of porous materials. The biggest advantage of these techniques is that they deliver an optical image of the samples.

Diffraction Techniques

Experiments using elastic X-ray and neutron scattering have turned out to be an invaluable tool for the characterization of various types of porous materials, providing quantitative parameters such as the pore size, surface area, and pore volume. In addition, diffraction techniques allow the determination of the shape and, in particular, the spatial distribution of the pores, for both highly ordered arrays such as in MCM-41 and a more disordered arrangement of pores, for instance in activated carbons. In the early 1900s, Max von Laue, W. L. Bragg, W. H. Bragg, and others laid the groundwork for X-ray crystallography, which has become a powerful method of visualizing complex inorganic and organic crystalline materials.In spite of the variety of different diffraction techniques currently available, they all are based on the same physical phenomenon, namely the scattering of X-rays and neutrons by the atoms through their electrons or nuclei, respectively. In the Fraunhofer approximation, the interaction of X-rays with electrons leads to the superposition of the coherent scattering of the basic scattering centers (atoms, molecules, or pores), without changing the energy of the incoming X-rays ("elastic scattering").

The resulting coherent scattering pattern is directly related to the mutual position, size, and scattering power of these scattering units. For almost all types of nanoporous materials, no single crystal diffraction datum is obtainable, not even for the crystalline zeolites. Therefore, most of the most prominent types of nanoporous materials (zeolites, highly ordered mesoporous materials such as MCM-41 or SBA-type materials, and porous carbons) are studied in diffraction experiments as polycrystalline powders. In this case, the powder diffraction raw data are obtained as 1D plots of the coherent scattering intensity versus the scattering angle 2θ. Only in the case of thin porous films with an oriented alignment of the pores relative to the substrate 2D diffraction can patterns be obtained. The main problem in analyzing scattering patterns of any kind of polycrystalline nanoporous materials lies in the extraction of a maximum of structural information: the highly advanced crystallographic strategies for the interpretation of 3D diffraction patterns of single crystals are not applicable, requiring different evaluation approaches. Depending on the length scale of the pore size, two methods can be distinguished regarding both the experimental realization and the theoretical treatment. Basically, the characteristic length scale d describing the pore system and the corresponding diffraction angle 2θ are related by the Bragg equation, $s = 1/d = (2\sin\theta)/\lambda$ where λ is the wavelength and s is the corresponding scattering vector.

Zeolites and other crystalline porous materials with pore sizes below the nanometer scale, which are obtained as polycrystalline powders, are investigated by wide-angle scattering (WAS) techniques, for instance by standard powder diffractometers, which are usually equipped with a 1D detector. The WAS scattering from a polycrystalline porous material is given by:

$$I(|\vec{s}|) = \left\langle |Z(\vec{s})|^2 |F(\vec{s})|^2 \right\rangle_\omega$$

where $Z(\vec{s})$ represents the 3D spatial distribution of the unit cell ("lattice factor"), and the electron distribution within the unit cell is described by the "atomic form factor" $F(\vec{s})$. The main difficulty in the interpretation of diffraction patterns of polycrystalline material arises from the spatial average $\langle \dots \rangle_\omega$, which leads to a loss of information compared with single crystals. Once the crystal structure has been solved, the pore size and shape are obtained from the atom coordinates.

Substantial progress has been achieved in the structure determination of zeolites based on powder diffraction data, which may comprise up to 200 reflection peaks in the wide-angle diffraction region. For instance, in the FOCUS software developed by Grosse-Kunstleve and co-workers, an automatic Fourier recycling algorithm is combined with a specialized topology (framework) search specific to zeolites. In essence, this algorithm takes into account chemical information on the composition and structure that is common to all zeolite frameworks. For instance, the evaluation algorithm includes the feature that these open-framework structures can be described as 3D four-connected nets of tetrahedrally coordinated atoms. Recently this concept was applied successfully to various types of zeolites, such as ZSM-5 and cloverite.

Compared with zeolites, nanoporous materials with pores sizes above 1 nm are characterized by a more disordered spatial distribution of the pores and are studied by so-called small-angle scattering (SAS), with the use of either X-rays (SAXS) or neutrons (SANS). The evaluation of SAS data for porous materials is usually based on the approximation that the material can be regarded as a so-called two-phase system: in the case of X-ray scattering, basic theoretical considerations show that the SAS of such materials arises from the scattering at the void-solid interface and is related to the scattering contrast $(\delta_1 - \delta_2)^2$, where δ_1 is the average electron density of the voids $(\delta_1 = 0)$ and δ_2 represents the average electron density of the solid. The SAS of a sufficiently ordered nanoporous material with pores of a distinct shape is also given by equation $I(|\vec{s}|) = \left\langle |Z(\vec{s})|^2 |F(\vec{s})|^2 \right\rangle_\omega$, where the lattice factor Z has the same meaning as in wide-angle scattering, and the form factor F here corresponds to the shape of the mesopore/solid rather than the electron density of single atoms. Polycrystalline mesoporous materials with a distinct pore shape and a well-defined 3D or 2D alignment, realized in silicas of the MCM or SBA families, give rise to SAS patterns with a characteristic sequence of reflections ("peaks"). The pore structures of even the most highly ordered mesoporous materials show a considerably lower order and symmetry compared with zeolites, leading to a much smaller number of possible mesostructures and corresponding SAS patterns. Therefore, the SAS patterns of this limited number of possible regular mesopore structures and space groups can serve as "ngerprints," allowing an almost unambiguous assignment of a certain mesopore lattice structure, if a sufficient number of refl ection peaks are obtained. In combination with TEM, SAXS experiments turned out to be a powerful technique for determining the alignment and structure in a variety of mesoporous MCM- and SBAtype materials and other structures. As the main information, SAXS provides the crystallographic space group of the mesopore arrangement and the corresponding lattice parameter. While the 3D alignment of the mesopores can be obtained from the bare SAS peak positions, the determination of mesopore sizes from singlepeak analyses involves substantial uncertainties, because both the peak profiles and intensities can be substantially superimposed by various factors such as smearing, incoherent background scattering, the arrangement of the mesopores and background fluctuations, and the presence of additional intrawall micropores. In particular, so far no satisfactory approach has been developed to quantitatively simulate the influence of 2D and 3D disorder on SAS. Even highly ordered mesopore systems such as MCM-41 show a certain 2D displacement of the cylindrical mesopores on the hexagonal lattice, which together with the polydispersity of the pores may result in a nonnegligible overlap of the peaks, thus severely aggravating a meaningful pore size analysis. In recent studies, the SAS of MCM-41 and SBA-15 silicas was modeled and compared with experimental data; however, the peak analyses did not take into account the impacts mentioned above. Figure shows typical SAXS patterns of three representative silicas exhibiting different degrees of structural order and regularity. Curve D in figure is the SAXS curve of SBA-2 silica, which has a hexagonal, close-packed structure.

Aside from the mesoporous materials with a high degree of order in terms of a regular spatial distribution and a high uniformity of pore size and shape, SAS is also applied to mesopores showing a certain disorder in the mesopore alignment and a pronounced polydispersity of pore sizes, which leads to SAS patterns with only few broad reflections. Smarsly et al. showed that the SAS of this type of material, possessing only a "liquidlike" local ordering of a system of polydisperse mesopores, can be reasonably approximated with the use of hard-sphere structure factors such as the Percus-Yevick approach. Curve C in figure represents the SAXS curve of a mesoporous silica obtained from PS-b-PEO templates ("SE1010 silica"). It is noteworthy that the SAXS data of both SBA-2 and SE1010 silica obey the Porod law,

$$\lim_{s\to\infty} s^4 I(s) = \frac{k}{2\pi^3 l_p}$$

which is theoretically predicted for a two-phase system, where l p is the so-called Porod length and k is the Porod invariant. Curve B in figure shows the SAXS curve of a porous silica with a disordered pore structure, templated with poly(ethylene oxide), also exhibiting the Porod asymptote at large values of the scattering vectors. It is obvious from figure, especially from curves B and C, that different approaches are in pore size and shape and an irregular spatial distribution of the micropores. Various evaluation methods have been used to determine the micropore size from SAXS. Whereas in certain studies the evaluation is based on a dilute system of polydisperse spheres, the application of the CLD concept has shown that the micropores in carbons are more consistent with an accute, needle-like shape. In addition, recent detailed SAXS studies indicate that the apparent fractal structures reported for certain microporous carbons are due to a misinterpretation of SAS data. Instead, Ruland and others have shown that the CLD concept provides an appropriate method for elucidating the microporous structure of carbons both qualitatively and quantitatively. Based on this method, the changes in the porosity that occur with typical physicochemical processing such as thermal treatment could be described by various structural parameters such as the average pore size, wall thickness, and angularity of the micropores. For example, SAS experiments revealed that the thermal treatment of carbons above ca. 1200 °C leads to an increase in the average pore size. In a recent study, the changes in the void-solid microstructure due to these treatments were studied by a combination of SAXS and WAXS.

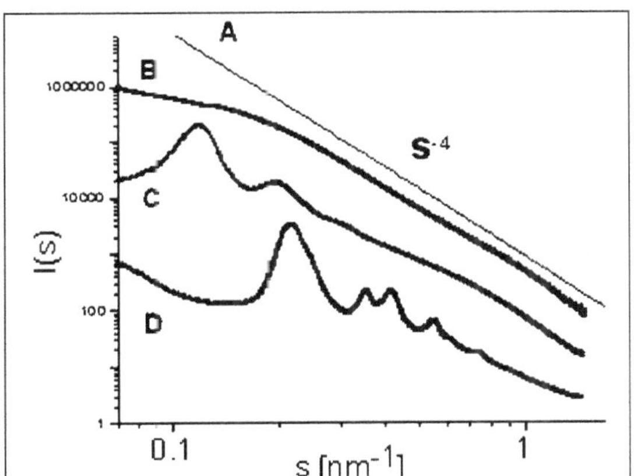

SAXS curves of various porous silicas. (A) Theoretical Porod asymptote for two-phase systems. (B) Poroussilica obtained from a PEO template. (C) Mesoporous SE1010 silica. (D) SBA-2.

Positron Annihilation

In the past 20 years, positron annihilation lifetime spectroscopy (PALS) has been developed into a powerful tool for the detection and quantification of defects on the atomic scale in various types of solids. PALS is sensitive to different kinds of defects, such as dislocations and vacancies in metals or crystals, grain boundaries, and voids and pores. Similar to scattering techniques, PALS is a noninvasive technique and thereby allows the detection of inaccessible pores. In the area of micro- and mesoporous materials, PALS is predominantly applied to porous polymers and thin porous lms. PALS is based on the decay of positrons into two *f* photons ("annihilation") and has been described in various publications. With Na as the radioactive source, the formation of positrons,

C5 by radioactive decay is accompanied by the simultaneous emergence of a γ-quantum of 1.273 MeV, which defines the starting signal of the positron lifetime measurement. Entering the sample, the positrons lose their high energy by inelastic collisions with electrons. These "thermalized" positrons have energies on the order of a few meV, form positroniums (Ps, the electron-positron bound state), and diffuse through the solid until annihilation after their specific lifetime in the solid, which is measured as the time difference between the creation of the 1.273-MeV γ -quantum and the annihilation radiation (two 511-keV γ rays). The natural lifetime of Ps of 142 ns is reduced by annihilation with electrons during collisions. The lifetimes, the inverse of the annihilation rates, become longer when a positron or positronium is localized at spaces with lower electron density such as voids.

Thus, positrons can be used as a probe to investigate average sizes of the free volume, size distribution, and the free volume concentration by measuring their lifetimes.The raw data of PALS are plots of the annihilation radiation signal as a function of time. In the case of defect concentration porosities that are not too high, there will be at least two or three lifetime components T_i in the spectra, which are usually analyzed as a sum of exponentials after background subtraction. Furthermore, the spectra can be deconvoluted with CONTIN software. Basically, PALS allows the calculation of both the porosity and the pore size distribution. Since the determination of pore size distributions is based on the trapping of positrons in voids of varying size, PALS was reported to lead to a single, average lifetime in the case of interconnected pore systems. The fundamental problem in the evaluation of PALS data lies with the exact relationship between the intensity of the long-lived components of a PALS spectrum and the concentration of the voids of a certain size, which is still a matter of intense research. For nonperiodic solids, a frequently applied model was developed by Tao and Eldrup. The extention of PALS to voids larger than 1 nm causes further uncertainties and is subject to ongoing research. PALS has been successfully applied to the investigation of porosity in various types of porous materials, such as thin siliceous films, cement, porous silicon, zeolites, and mesoporous silicas such as vycor glass.

Mercury Porosimetry

In mercury porosimetry (MP), gas is evacuated at elevated temperatures and low pressures from the sample placed inside a pressure chamber. Subsequently the sample is immersed in mercury and an external pressure is applied to force mercury into the sample. By gradually increasing the pressure and monitoring the incremental volume of mercury intruded for each new applied pressure, the pore size distribution of the sample can be estimated in terms of the volume of the pores intruded for a given diameter D. Mercury is a nonwetting fluid at room temperature for most porous materials of technological interest. The evaluation of pore sizes from MP is based on the Washburn equation, quantifying the pressure p required to force a nonwetting fluid into a circular cross-sectional capillary of diameter D by $p = (4\gamma \cos\theta)/D$ where γ is the surface tension of mercury and θ is the contact angle on the surface of a solid sample. MP allows the determination of average pore sizes (and their distribution) between 3 nm and 200 nm and is therefore inappropriate for microporous materials, but more suitable for pore sizes above ca. 30 nm than nitrogen sorption. As further parameters, the total pore volume V_{tot} is accessible from the total intruded volume of mercury at the highest pressure determined, and the total pore surface S is calculated from $S=(1/\gamma|\cos\theta)\int_0^{V_{tot}} p\,dV$. Because of inherent shortcuts MP is not used as frequently as gas sorption techniques for samples with mesopore sizes below 20 nm. During the measurement, high pressures used to force mercury into small pores may compress the sample. Damage or compression of highly porous silica has been reported previously. In addition, mercury porosimetry overestimates the volume of the smallest pores in the case of ink-bottle-shaped pores, because the intrusion of mercury into the larger pores is determined by the small openings. Moreover, it has to be pointed out that pore size distributions by the Washburn equation are not a geometrical relationship, but a physical characteristic of a porous medium, because MP is based on transport and relaxation phenomena.

General Strategies for Producing Porous Materials of any Kind

The most successful way to produce all sorts of nanoporous materials is the templating method, shown schematically in figure.

An organic (or sometimes inorganic) compound acts as a place holder what later becomes a void space in nanoporous materials. The templating concept obviously allows control, mainly of the criterion of pore size but also pore shape. At first, a suitable template structure has to be provided. By “suitable” we mean that this template structure must constantly be compatible with solvents involved in the process and the final network materials. Macrophase separation must be avoided at all times. the creation of a spheres (5) or assemblies of spheres (6). These units are surrounded or bonded to the network precursor systems in the actual templating step (b). The created pore resembles the size, shape, and even functional characteristics of the template in an ideal case. Finally, the template is removed. Therefore, it is possible to tune the properties and most importantly the size and interconnectivity of the pore and pore system in the order (1)–(6).

Figure shows schematic representation of the approach to obtaining nanoporous materials by templating strategies. (a) A range of applicable templates beginning with single molecules (1) or amphiphiles (2), followed by assemblies of amphiphiles as micelles (3) or lyotropic phases (4), followed by polymeric porous material is always connected to the creation of a high surface area. In other words, the interface energy between the template structure and the network material despite this huge interface area has to be smaller than kT. Otherwise the material can rearrange into more stable, nonporous phases.

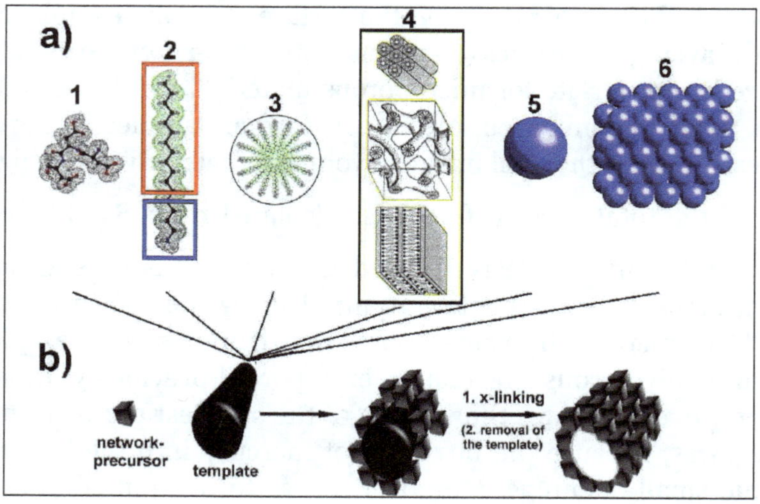

The template structure can have sizes on many different length scales. The smallest pores are achieved for molecules as templates. The pores created have a shape and size similar to those of the molecules used. The pores can be spatially uncorrelated with each other. In this case one speaks about molecular imprinting. Molecular imprinting is even successful for organic materials used to build networks and is therefore very often used for imprinting biomolecules to a matrix for biochemical applications. Highly ordered materials with pores that have a correlation with each other are found for zeolitic materials. The most prominent examples for zeolites are found in silica chemistry, and there are some naturally occurring compounds. Here it is enough to mention that primarily amines and ammonium cations as templates determine the formation of pores in the range of 1 nm to 1.5 nm. The materials have high symmetry and a crystalline framework.

Larger pores can now be achieved by assemblies of molecules forming templates. Here assemblies of amphiphiles are ideal because of their selforganizing properties and their capability for lowering interface energy. An amphiphilic molecule typically has one hydrophilic and one hydrophobic part, which tend to lead to microphase separation. The resulting template structures are micelles or ultimately lyotropic phases with different possible symmetries. Materials with pores on the order of 2–50 nm (so called mesoporous materials) result, with a sometimes surprisingly high order. Even larger pores are obtained when polymer spheres or colloidal crystals of these sphere are used as templates.

Microporous Material

A microporous material is a material containing pores with diameters less than 2 nm. Examples of microporous materials include zeolites and metal-organic frameworks.

Porous materials are classified into several kinds by their size. The recommendations of a panel convened by the International Union of Pure and Applied Chemistry (IUPAC) are:

- Microporous materials have pore diameters of less than 2 nm.
- Mesoporous materials have pore diameters between 2 nm and 50 nm.
- Macroporous materials have pore diameters of greater than 50 nm.

Micropores may be defined differently in other contexts. For example, in the context of porous aggregations such as soil, micropores are defined as cavities with sizes less than 30 μm.

Uses in Laboratories

Microporous materials are often used in laboratory environments to facilitate contaminant-free exchange of gases. Mold spores, bacteria, and other airborne contaminants will become trapped, while allowing gases to pass through the material. This allows for a sterile environment within the contained area.

Use in Medicine

Albupore microporous surgical tape.

Microporous adhesive tape is a surgical tape used to hold wound dressings and bandages in place, introduced in 1959 by 3M with the trade name Micropore. It can be used to hold gauze padding over small wounds, usually as a temporary measure until a suitable dressing is applied. The Steri-Strip was derived from Microporous surgical tape.

Other Uses

Microporous tape is used by some professional extreme yo-yoers to wrap around their fingers and prevent string burn or irritation.

Rock climbers use microporous tape to wrap their hands in 'tape gloves', a means of protecting the skin from rock abrasion when jamming hands into cracks as a means of ascent (crack climbing, as opposed to face climbing - gripping holds on the face of the rock).

Microporous media used in large format printing applications normally with a pigment based ink to maintain colour balance and life expectancy of the resultant printed image.

Microporous tape is also used by some film and TV sound recordists to affix small radio microphones to actors' skin.

Microporous material is also used as high performance insulation material used from home applications up to metal furnaces requiring material that can withstand more than 100 °C.

Mesoporous Material

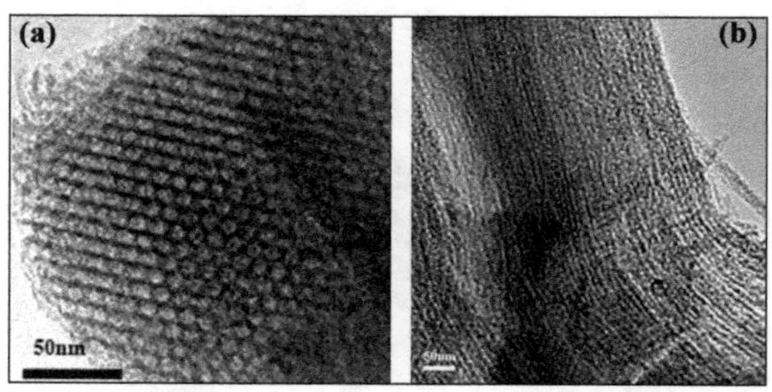

Electron microscopy images of nitrogen-containing ordered mesoporous carbon (N-OMC) taken (a) along and (b) perpendicular to the channel direction.

A mesoporous material is a material containing pores with diameters between 2 and 50 nm, according to IUPAC nomenclature. For comparison, IUPAC defines microporous material as a material having pores smaller than 2 nm in diameter and macroporous material as a material having pores larger than 50 nm in diameter.

Typical mesoporous materials include some kinds of silica and alumina that have similarly-sized mesopores. Mesoporous oxides of niobium, tantalum, titanium, zirconium, cerium and tin have also been reported. However, the flagship of mesoporous materials is mesoporous carbon, which has direct applications in energy storage devices. Mesoporous carbon has porosity within the mesopore range and this significantly increases the specific surface area. Another very common mesoporous material is Activated Carbon which is typically composed of a carbon framework with both mesoporosity and microporosity depending on the conditions under which it was synthesized.

According to IUPAC, a mesoporous material can be disordered or ordered in a mesostructure. In crystalline inorganic materials, mesoporous structure noticeably limits the number of lattice units, and this significantly changes the solid-state chemistry. For example, the battery performance of mesoporous electroactive materials is significantly different from that of their bulk structure.

A procedure for producing mesoporous materials (silica) was patented around 1970, and methods based on the Stöber process from 1968 were still in use in 2015. It went almost unnoticed and was reproduced in 1997. Mesoporous silica nanoparticles (MSNs) were independently synthesized in 1990 by researchers in Japan. They were later produced also at Mobil Corporation laboratories and named Mobil Crystalline Materials, or MCM-41. The initial synthetic methods did not permit to control the quality of the secondary level of porosity generated. It was only by employing quaternary ammonium cations and silanization agents during the synthesis that the materials exhibited a true level of hierarchical porosity and enhanced textural properties.

Since then, research in this field has steadily grown. Notable examples of prospective industrial applications are catalysis, sorption, gas sensing, ion exchange, optics, and photovoltaics.

It should be taken into account that this mesoporosity refers to the classification of nanoscale porosity, and mesopores may be defined differently in other contexts; for example, mesopores are defined as cavities with sizes in the range 30 μm–75 μm in the context of porous aggregations such as soil.

Macroporous Materials

Macropores are defined as cavities that are larger than 75 μm. Functionally, pores of this size host preferential soil solution flow and rapid transport of solutes and colloids. Macropores increase the hydraulic conductivity of soil, allowing water to infiltrate and drain quickly, and shallow groundwater to move relatively rapidly via lateral flow. In soil, macropores are created by plant roots, soil cracks, soil fauna, and by aggregation of soil particles into peds.

Macropores may be defined differently in other contexts. Within the context of porous solids (i.e., not porous aggregations such as soil), colloid and surface chemists define macropores as cavities that are larger than 50 nm.

Opals and Inverted Opals

It is possible to cast inorganic materials by suitable templating techniques with high precision. Templates originating from the self-assembly of amphiphilic molecules have been discussed, and it was pointed out that colloidal objects or colloidal crystals can also be used as templates to obtain pores of a higher length scale. It is known that colloidal particles such as silica spheres, latex spheres, or inorganic crystals (such as gold or semiconductors), as soon as their size is fairly monodisperse, tend to pack into ordered assemblies called colloidal crystals.

Figure shows the colloidal crystal formed by negatively charged 280-nm latex spheres. As can be seen from this image, the dense packing of 280-nm spheres creates regular voids in the size range of 70 nm. Therefore, colloidal crystals obtained by spheres or sphere templating techniques belong to the family of nanoporous materials.

A discussion of the physics behind the formation of colloidal crystals is beyond the scope of this review article and should be taken from the literature. However, common techniques for obtaining colloidal crystals of spheres are centrifugation, electrophoresis, and evaporation methods. Concerning the last method, which is the convective self assembly process, should be mentioned explicitly because it is a very good and easy method for obtaining opal structures of spheres. Very high ordered opal structures can be obtained, but the symmetry seems to be restricted to face-centered cubic (fcc), as expected from theoretical calculations. However, in rare cases other structures, such as (bcc), (hcp), and (rchp), could be observed.

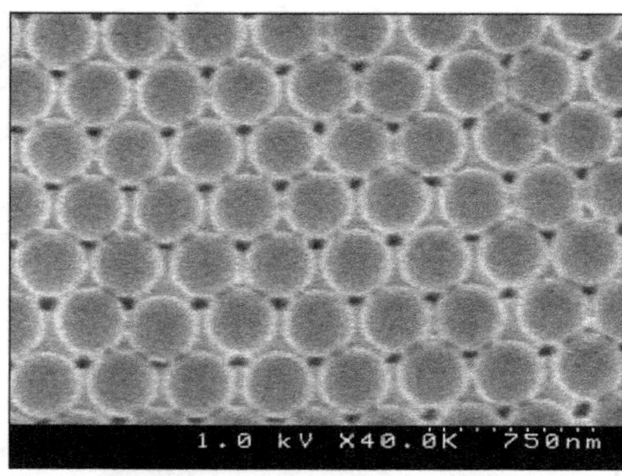

SEM image of a nanoporous material formed by a colloidal crystal of latex spheres.

The spheres in the opal structures discussed so far are held together just by relatively weak forces. For later applications cracking or disrupting the structure is always a problem. By sintering or thermal annealing of silica spheres, or heating above the glass point for organic polymer spheres, it is possible to strengthen the material.

Much attention has recently been paid to the transformation of these opal structures into inverted opals by templating techniques. Because in these cases voids, sometimes on the order of several micrometers, are produced, just the basic principles are described in this review about nanoporous materials. A more detailed description can be found elsewhere. The interstices of the opals are filled with liquid or gaseous (CVD) precursors, which react to form a new network. Afterward, the sphere template is removed by calcination for polymer spheres or by HF etching for silica spheres. In this way, a large variety of inverted opals ranging from inorganic oxides such as titania, to metals, carbon, or semiconductors have been prepared.

Photonic Properties

All of the opal materials discussed show, because of their architecture, a characteristic modulation of dielectric constant in 3D space. This is the major requirement for so-called photonic materials. In simple terms, a modulation of the dielectric constant or the order of light leads to Bragg diffraction for specific light frequencies. This means that certain light frequencies cannot propagate through the opals. These frequencies where light is reflected are also called stop-bands. Further requirements for photonic materials are a high refractive index contrast and a high order in the material.

Although the combination of Bragg's law and Snell's law gives a good approximation for the peak position and the dependence on the lattice constant, filling fraction, and refractive index contrast, a more detailed description is required:

$$\lambda_{C(hkl)} = 2.d_{(hkl)} \cdot \sqrt{\langle \varepsilon \rangle - \sin^2 \theta}$$

where $\lambda_{C(hkl)}$ is the wavelength for the stop-band, $d_{(hkl)}$ is the lattice constant, $\langle \varepsilon \rangle$ is the effective refractive index, and θ is the angle.

It is beyond the scope of this review article to give a detailed description and derivation of the theories behind photonic properties, but to understand the basic principles it might be interesting to mention that photonic materials are often seen as "semiconductors for light." It is believed that the flow of light can be controlled with photonic materials in a way similar to that in which a semiconductor controls the flow of electrons. Indeed, the theories behind the behavior of light or electrons in solids are somewhat similar because both systems can be described as waves. However, because electrons are fermions and photons are bosons, there are also important differences. The 3D modulation in refractive index has consequences for photons similar to those the 3D modulation of electrostatic potential in a solid material has for electrons. Depending on the propagation direction of light inside the material, bands of certain energies are created that allow light of these frequencies to propagate. But more important, for some energies there are no states in the material, which means that light of this energy is not allowed inside the material. It is reflected, which is detectable by UV/Vis spectroscopy.

Figure shows the correlation between the light-band structure inside a photonic crystal and the UV/vis spectrum obtained. As shown, there is a certain energy region (in this case above 1.0 eV) where no states for photons exist. As a consequence, no photons are transmitted through the material.

Table: Opals, semiconductors for light.

Semiconductors and electrons	Opals and photons
3D distribution of atoms	3D distribution of scattering centers
Periodic electrostatic potential	Periodic dielectric constant
Description by Schrödinger	Description by wave

equation (scalar)	equation (vectorial)
$\frac{\bar{h}}{2m}\frac{d^2\psi}{dx^2}+(E-V)\psi=0$	$\frac{d^2E}{dx^2}+(n\omega)^2E=0$
Fermions	Bosons
Creates bands with allowed	Creates bands with allowed
electronic states (energies) and band gaps	electronic states (energies) and band gaps

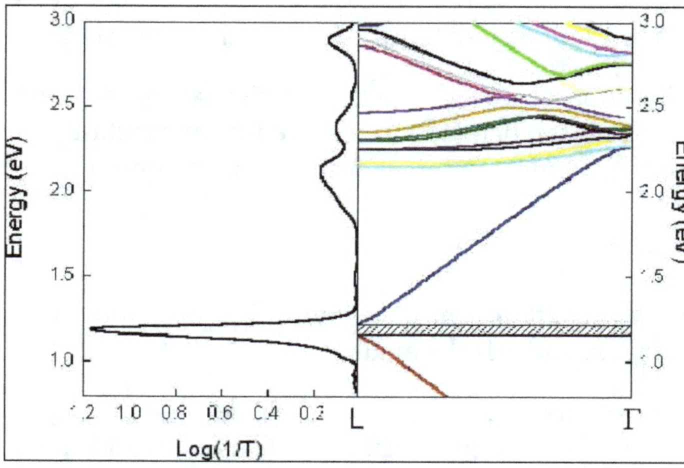

The band structure for allowed light states (right) is shown by the colored lines, depending on the propagation angle inside the opal (x axis). The left side of the figure shows the corresponding UV/vis spectrum.

Many, many challenges and exciting developments can be expected for this new area of nanoporous materials. Current interests focus on obtaining special structure types of opals, such as the diamond structure; improving the refractive index contrast; or preparing microengineered arrays of opals. Further applications of these materials lie in the areas of waveguides, lters, microcavity lasers, or optical chips.

Example of Nanoporous Solids

Activated Carbons

Activated carbons are key materials in technological applications of multidisciplinary fields (e.g. adsorption, separation, and catalytic processes). The extensive use of these materials results from the combination of a well-developed pore network (micropores or micro + mesopores) along with the presence of heteroatoms (e.g. oxygen, nitrogen, and sulfur). The large scale production of nanoporous carbons is a well-established process since the first patents date from the beginning of the twentieth century. Conventional activation methodologies are divided between physical, using steam or CO_2, and chemical, being KOH, $ZnCl_2$, and H_3PO_4 the most commonly reported oxidizing agents. Due to the panoply of operational parameters that can be changed or added in the production of activated carbons, there is still room for R&D.

Carbon is the main element of the entire world's flora, and its atomic structure allows unique bonding possibilities leading to various structures with distinct properties. As consequence, carbon materials are used for so long that it is hard to determine the very first moment.

At the beginning, carbon was used in the form of charcoal that consists of carbonized wood, coal or partially devolatilized coals, and also carbon black obtained by incomplete burning of vegetable oil. After centuries of use of these man-made forms of carbon, the discovery of the processes to perform the activation of charcoal allowed to greatly improve the properties and performances of carbon materials. These new materials where pores (voids) are enclosed by carbon atoms are known as activated carbons or active carbons and, more recently, also named as nanoporous carbons.

The first large-scale application in gas phase took place in 1854, with the installation of carbon filters in London sewage ventilation systems. In 1872, carbon filters were also used in the masks of the chemical industry workers for preventing the inhalation of mercury vapors.

Besides thousands of years of history, and the already large range of applications for charcoals, it was only in the beginning of the twentieth century that one of the most outstanding abilities of carbon materials was explored: The possibility of enclosing a huge porosity into the carbon material structure. This "revolution" was due to the discoveries of Raphael von Ostrejko that, between 1900 and 1903, patented two different methods for the industrial activation of charcoal and the production of activated carbon materials. This set of patents mentions the bases of the chemical activation process (carbonization of lignocellulosic materials with metallic chlorides) and of the thermal or physical activation (slight gasification of chars with water vapor or carbon dioxide at elevated temperatures) and also the specific equipment for thereof. The factory for full-scale activated carbon production was built by that time in Ratibor, now named Raciborz (Poland), and still operates, being so the world's oldest industrial activated carbon manufacturing plant.

Although sugar industry was the first to apply activated carbons, in this case for whitening purposes, the starting point for the great development in the production and application of activated carbons was undoubtedly World War I, when these materials were used in gas masks. The production and search for new activated carbons has been boosted decade after decade due to their fundamental role in various technological applications which are related to, namely, restricted environmental regulations, recovery of valuable chemical compounds, and catalysts support.

Nowadays, the driving forces for the research in nanoporous carbons are related with the properties of the most recent carbon materials: fullerenes, carbon nanotubes and graphene. However, the excellent properties of these novel carbon forms also fostered the interest in the traditional porous carbons and, in the recent years, a considerable number of studies searching for new synthetic approaches have been published. The main objective is the preparation of highly porous materials with controlled porosity, and often also with tuned surface chemistry, to present enhanced behavior as, for example, electrode materials for supercapacitors.

Properties of Nanoporous Carbons

Describing nanoporous carbon materials as carbon structures with enclosed porosity is correct but it is somewhat incomplete since properties, as for example, surface chemistry, chemical composition, morphology or electrical conductivity are also essential to understand their potential and performance in a given application.

Regarding the physical properties of nanoporous carbon materials, porosity is undoubtedly the most important one. IUPAC defines nanopores as pores with apertures up to around 100 nm and, according to their size, nanopores are classified as follows:

- Micropores—pores with width less than 2 nm.
- Mesopores—pores with width between 2 and 50 nm.
- Macropores—pores with width greater than 50 nm.

Due to their relevance for the adsorption process, micropores can be further distinguished between narrow micropores (i.e. ultramicropores) with width <0.7 nm and wide micropores (i.e. supermicropores) with width between 0.7 and 2 nm.

The limits for this pore classification were suggested by nitrogen adsorption-desorption data at −196 °C and rely on the fact that each pore size range corresponds to different pore filling mechanisms disclosed by the isotherm profile. Micropore filling is regarded as a primary physisorption process and occurs at low relative pressures. Firstly, at very low relative pressure, adsorption occurs in the ultramicropores where due to the close vicinity of pore walls there is an overlapping of the adsorption forces favoring the enhancement of the adsorbent-adsorbate interaction, this process is termed "primary micropore filling." After this initial filling in the most energetic sites, the cooperative adsorbate-adsorbate interactions inside the supermicropores become more relevant (secondary process) occurring up to relative pressures ≈ 0.15. In mesopores, the physisorption comprises three stages: monolayer adsorption (i.e. all adsorbate molecules are in contact with the surface of the adsorbent) followed by consecutive adsorbate layers (multilayer adsorption) and finishing with capillary condensation. The adsorption onto mesopores commonly presents a hysteresis loop. The macropores are so wide that do not allow capillary condensation since behave as an open space (cannot be characterized by nitrogen adsorption whose maximum relative pressure occurs near the unity).

In nanoporous carbons, the porosity results from the voids between the randomly cross-linked graphite-like crystallites that constitute the structure of the carbon skeleton, from the consumption of less organized carbonaceous matter and also from the removal of reactive carbon atoms in the crystallite during activation process. Surface area and pore size distribution (PSD) are the physical properties that will greatly influence the performance of nanoporous carbons in a given process. These parameters are commonly obtained from the analysis of nitrogen adsorption isotherm at −196 °C. The surface area (apparent surface area for microporous materials) is commonly obtained applying the Brunauer-Emmett-Teller (BET) equation. Regarding the PSD, there are various mathematic models available, being those using the Non-Local Density Functional Theory (NLDFT) the most advanced ones. The use of nitrogen as adsorptive at −196 °C has however diffusional limitations in ultamicroporous carbon materials; thus, for an accurate characterization of the micropore network, carbon dioxide adsorption at 0 °C has been repeatedly proposed as a better alternative to nitrogen. Besides the limitation of nitrogen for the characterization of ultramicropores, it is known that, due to its quadrupole moment, nitrogen molecule orientation during adsorption will be dependent on the surface chemistry of the adsorbent. To overcome this issue, other adsorptive gases like argon or krypton can be used.

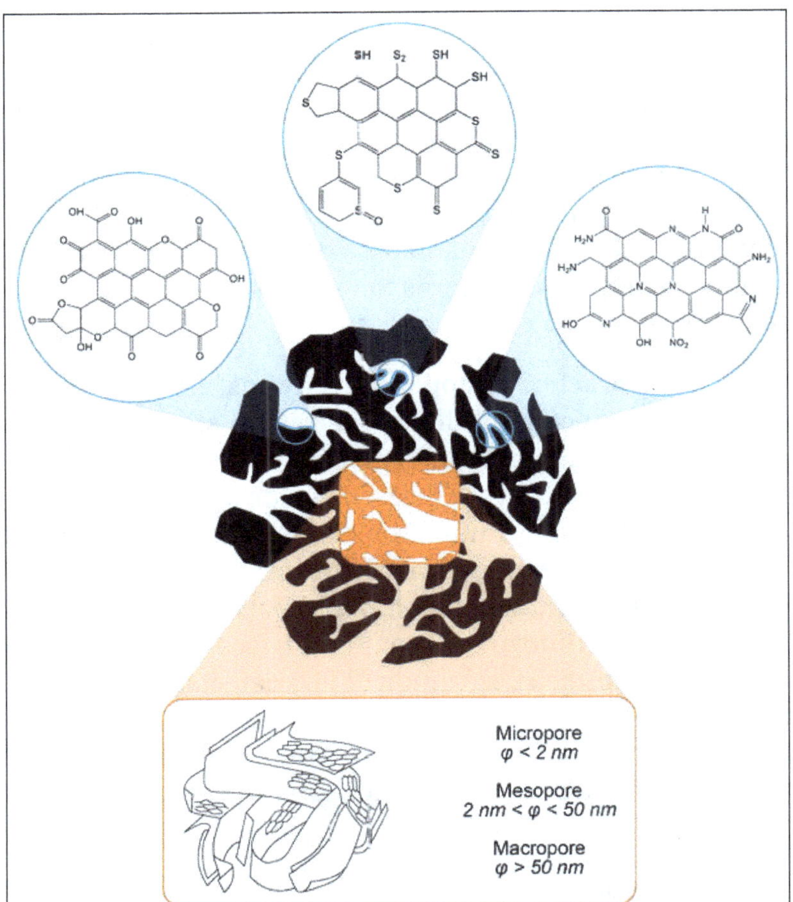

Schematic representation of pore structure (bottom rectangle) and surface chemistry—oxygen, nitrogen, and sulfur groups (top circles)—in nanoporous carbons.

Carbon atoms are the major element of nanoporous carbon skeleton but the elemental composition of these materials also includes hydrogen and oxygen and, depending on the precursor, preparation route and post-synthesis functionalization may also have nitrogen, sulfur or phosphorus containing groups. These heteroatoms are mainly located at the edges of the basal planes due to the presence of unsaturated carbon atoms that are highly reactive. Figure summarizes the most relevant oxygen, nitrogen, and sulfur-containing surface groups that may be present in the surface of nanoporous carbon materials, being important to mention that oxygen surface groups are, by far, the most common ones. These materials may also contain inorganic matter, sometimes attaining 20% (wt.), being this ash content mainly inherited from the carbon precursor.

The elemental composition and type of surface groups of a nanoporous carbon influence its performance in both gaseous and liquid phase processes, due to specific interactions with the adsorptive and also the solvent in the case of adsorption from solution. Properties like hydrophobicity/hydrophilicity or acid/base behavior are highly dependent on the surface chemistry of these materials. Carbons are in general hydrophobic, but the presence of increasing amounts of oxygen surface functionalities favors the adsorption of water molecules due to the formation of hydrogen bonds and, consequently, its wettability. This ability of porous carbon materials may be advantageous, or not, depending on the desired application. For example, in the impregnation of carbon supports with catalysts in aqueous phase, a high wettability will increase the impregnation degree; but in the adsorption of organic compounds from diluted aqueous solutions, higher wettability may led to the formation of water molecule clusters that hinder the diffusion of the target compounds towards the adsorption active sites (the same being valid for gas steams purification).

Regarding acid/base character, nanoporous carbons are considered amphoteric materials due to the presence of both acid and basic sites in their surface. Thus, depending on the amount and strength of all the surface groups, the materials may present net acid, basic or neutral surfaces. The surface chemistry of nanoporous carbons can be assessed through numerous techniques, and the best way to get a good characterization is to employ complementary techniques and combine the analysis of the results. For example, Boehm and potentiometric titrations provide qualitative and quantitative information on the surface of the nanoporous carbons, while temperature programmed desorption (TPD) detects more oxygen groups than Boehm titration, although with less quantitative information. On the other hand, X-ray photoelectron spectroscopy (XPS) and diffuse reflectance infrared spectroscopy (DRIFT) provide qualitative information about the surface of the nanoporous carbons.

Other properties of nanoporous carbon materials that may play important roles in specific applications are the morphology (powdered (PAC) and granular (GAC) forms are the most common ones), electrical conductivity, hardness and density.

Nanoporous Carbon Synthesis

There are a large range of carbon-rich precursors namely from vegetable (e.g. wood, fruit shells, stones or peels) or petrochemical origin that can be successfully transformed into activated carbons. However, the economically viable large-scale production of activated carbons requests the use of raw materials that gather high availability and constancy with high density and hardness, low inorganic content and, last but not the least, low cost. Thus, commercially available activated carbons are mainly produced from coal (i.e. lignite, sub-bituminous, bituminous and anthracite), coconut shell charcoal and wood charcoal. The raw material is a key issue for activated carbon producers since the development of the porous network occurs by consumption of the (char)coal precursor and the preparation yields can reach values as low as 10% (wt.). Consequently, assuring the supply of high quality precursors at a cost-effective price is of paramount importance to control the quality of the final material and the production cost. In 2015, the global activated carbon market was valued by Markets and Markets at USD 4.74 billion corresponding to 2743.7 kton being expected to continue to growth, mainly driven by more stringent government regulations to assure human and environmental protection. (char)coal is the most important raw material for activated carbon producers but they are not the major (char)coal consumer. In fact, activated carbons are only a "by-product" of (char)coal production, which have the energy sector as major market (mainly for metallurgical, industrial and cooking fuel) that actually controls demand and prices. This reality, allied to the fact that most of raw materials came from the Asian continent, and that external factors (e.g. natural disasters) may compromise the quality, availability and cost of (char)coal, continues to drive the research for finding new, and more sustainable, raw materials and preparation routes for activated carbon production.

Conventional methods for activated carbon production request the use of high temperature kilns where, under a controlled atmosphere, the carbon-rich raw materials are transformed into nanoporous carbons. There are numerous procedures for the production of activated carbons, although the great majority of synthetic routes published and patents registered worldwide, including those allowing nowadays industrial production, fall in two major categories: the selective gasification of raw materials' carbon atoms (physical or thermal activation), and the co-carbonization of the precursor with oxidizing and/or dehydrating agents (chemical activation). The number of steps requested for the production of activated carbons depends on the characteristics of the raw material, morphological specifications of the final product, and on the type of activation. A general process flowsheet.

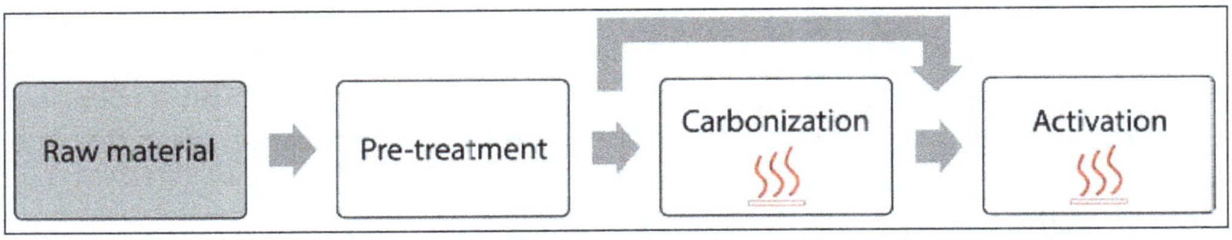

General flowsheet for conventional physical (thermal) and chemical activation processes.

The most commonly used pre-treatments are those aiming to obtain a given particle size or shape, washing steps for dirt removal and/or for the reduction of the inorganic content (i.e. acid washing) and pre-oxidation to prevent fluidization of coking coal during carbonization. The raw material can be crushed and sieved to obtain powdered or granular particles with specific dimensions. Milling, binder addition and briquetting are other common options when working with precursors presenting low volatile content (e.g. medium-high rank coals, volatiles <20–30% (wt.)). To assure an efficient diffusion of the activating agent to all the individual coal particles of the briquette, it is common to mill and subsequently briquette the particles to obtain granular materials with a well-developed network of transport pores (i.e. meso and macropores). Direct activation of granular coal with low volatile content will not allow an uniform activation of the granules due to the incipient network of transport pores.

Physical or thermal activation is generally made in two consecutive heating steps: carbonization under inert atmosphere (usually nitrogen) to devolatilize the raw material, followed by activation that consists in the partial gasification of the obtained char with oxidizing agents (i.e. steam, carbon dioxide or a mixture of both) leading to the formation of the porous network. While the carbonization usually occurs at temperatures between 400 and 600 °C, the gasification requests temperatures ranging from 800 to 1000 °C. Depending on the raw material, it is also possible to avoid the carbonization step and proceed directly to the thermal activation. Activation with CO_2 must occur in conditions that assure chemical control (slow activation rate—days) instead of diffusional control which is faster but leads to external particle burning and, consequently, to poor porosity development. The reactions of steam and carbon dioxide with carbon are endothermic, thus thermal activation needs external energy supply to maintain the requested high temperature. Oxygen (or air) is not commonly used as oxidizing agent since its reaction with carbon is highly exothermic and fast, being difficult to control and assure porosity development instead of particle consumption. Due to this and to the safety issues related to the temperature control, oxygen activation is scarcely used. However, low amounts of oxygen (or air) can be added during thermal activation with steam or carbon dioxide to help maintaining the high temperatures by reacting with the gases produced during activation (i.e. CO and H_2). This approach has the advantage of lowering the pressure of CO and H_2, both inhibiting gases for the activation, and increasing the partial pressure of the activating agent.

Chemical activation usually requests only one heating step: the raw material is mixed with an activating agent (e.g. $ZnCl_2$, H_3PO_4, KOH) and is further treated under controlled atmosphere at temperatures from 400 to 900 °C, depending on the selected activating agent. The solid product obtained need to be exhaustively washed with water to remove the chemicals and dried before storage. In this process, the mechanism of pore formation is dependent on the chemical agent: Zinc chloride promotes the removal of water molecules from the lignocellulosic structures of the raw material while phosphoric acid combines chemically with them. In none of these processes, the selective removal of carbon atoms occurs. With potassium hydroxide, the process is more complex since there is the disintegration of the structure followed by intercalation of the metallic potassium into the "graphitic" laminar structure, breaking down particles and limiting granular activated carbons synthesis. Simultaneously, there is also some gasification of carbon atoms due to reaction with CO_2 and H_2O, resulting from the redox reaction of carbon with potassium compounds. During carbonization (i.e. heat treatment under inert atmosphere), the lignocellulosic precursor losses volume by contraction, but when chemical activation is applied there is incorporation of the activating reagent inside the particles inhibiting the expected contraction, that is, the activating agent may behave as a template for the formation of the microporosity. In fact, in 2004, Molina-Sabio and Rodríguez-Reinoso published an enlightening research that allowed to conclude that the dehydration of the carbon precursor (i.e. peach stones) by $ZnCl_2$ occurs along with templating by the activating agent, thus justifying the porosity evolution with the amount of $ZnCl_2$ impregnated.

Chemical activation has advantages over physical process related with the use of a single heating step and lower temperatures, usually higher yields, shorter activation time (hours) and higher surface area and pore volumes. All these add-ons came with a price since this activation process is more corrosive (e.g. $ZnCl_2$ or KOH) than the physical one, and activating agents can be hazardous for the environment (i.e. $ZnCl_2$ and H_3PO_4). Also, regardless the activating agent, a time-consuming washing step is needed to remove chemicals from the newly formed pore network and, if possible, to recover the activating agent. Actually, $ZnCl_2$ use is declining due to the environmental issues associated with zinc residues, low recovery efficiencies, corrosion problems and presence of residual zinc in final carbon.

Zinc chlorine and phosphoric acid lead to higher activation yields (≈40%) compared to physical activation since the use of these compounds inhibit the formation of tar and other by-products formed in physical activation. In the case of potassium hydroxide, the more complex activation mechanism gathering K intercalation in carbon lattice and gasification allows to prepare activated carbon with a large range of yields depending on the precursor: From 80 and 70% to values, as low as, 20–10%. KOH allows obtaining superactivated carbons (over 3000 $m^2 g^{-1}$) when high KOH/precursor weight ratios (i.e. between 2 and 4) are used.

It is important to emphasize that the effect of experimental conditions on pore size distribution is a general trend; thus, for an in-depth knowledge of the effect of activating agent and experimental conditions on the porous characteristics of a given raw material, a comprehensive study is always needed.

The above-mentioned precursors and activating agents are those more representative for industrial scale production of activated carbons. However, regarding the production of specialty carbons (low volume processes) and for research works, the number of options increases exponentially. Regarding raw materials, hard biomass residues like shells and stones allow to obtain high quality PACs or GACs; phenol formaldehyde polymers yield high surface area porous carbons, while polyacrylonitrile (PAN), acrylonitrile textile or rayon are adequate to synthetize activated carbon fibers or cloths. When aiming nitrogen doped materials, for example for the adsorption of sulfur species, the use of polymers containing nitrogen (e.g. PAN) is a common synthetic route. In what concerns activating agents, besides $ZnCl_2$, H_3PO_4 and KOH, alkali hydroxides and salts (e.g. NaOH, K_2CO_3, Na_2CO_3) and metal chlorides (e.g. $AlCl_3$, $FeCl_3$, NH_4Cl, $CuCl_2$) can also be used.

At industrial scale, steam activation continues to be the most widely used method to produce activated carbons. The advantages of steam activation are related with the fact that does not request post-activation work-up, namely a final washing step, is less expensive and has less environmental constrains than the chemical activation. Besides, for precursors with less than 10% of ash, carbon materials with surface areas of 1000 $m^2 g^{-1}$ can be easily obtained with activation yields of 50%, thus allowing a good compromise between production cost and porosity development.

In general, activating methodologies produce activated carbons with wide pore size distributions, so the need for carbon materials with more regular porosity prompted the research on the synthesis of ordered mesoporous carbons (OMCs) by applying hard or soft templating approaches. The advantages of OMCs over conventional activated carbons are related with the ordered and hierarchical pore network, but their multi-step preparation procedures are time consuming, use hazardous chemicals to remove the inorganic templates and, consequently, have very low atomic economy and high production costs.

Conventional Activation of Gels and Chars Obtained by Novel Approaches

Classical preparation of activated carbons usually requests an initial thermochemical process (carbonization) to remove other elements than carbon and transform the raw material (i.e. coal or biomass) in a carbon-rich material (charcoal) that will be subsequently activated to improve the pore network. This thermochemical conversion is particularly important when using biomass whose carbon content usually is between 30 and 50% (wt.). However, the need for more specialized porous carbon materials, as well as the growing interest in developing more sustainable and greener processes, has prompted the research community to explore alternatives to the conventional carbonization step which, in some cases, also allow to make feasible the use of novel precursors.

In the category of alternative methods to obtain carbon-rich materials for further activation via conventional methodologies, processes as distinct as sol-gel polymerization reaction, hydrothermal carbonization (HTC) and acid-mediated carbonization (AMC) can be grouped. While sol-gel process involves the use of synthetic compounds—resorcinol and formaldehyde—as reagents for the polycondensation reaction in the presence of a catalyst, both HTC and AMC are mainly applied to transform biomass into a coal-like material.

The first resorcinol-formaldehyde organic gel was synthetized by Pekala in 1989 being the aqueous polycondensation performed under alkaline conditions. One of the major advantages of this process is its flexibility since the main properties of the gel can be tuned during the synthesis and drying process. The synthesis starts with the formation of the wet gel and, during this step, the most important parameters controlling the properties of the final gel are the precursors' concentration, the catalyst type and concentration and the time and temperature of curing. After drying, aerogels (supercritical drying), xerogels (subcritical drying), and cryogels (freeze drying) can be obtained. The drying procedure is one of the most important steps regarding the porous properties of the organic gel and is also crucial when the preparation of a thermally stable carbon gel is foreseen. The organic gels are mainly mesoporous materials thus if the preparation of a micro + mesoporous material is envisage a thermochemical step involving carbonization and/or activation must be considered. The first works focused on the carbonization and activation of organic gels were published during the 1990s, and nowadays, these methodologies continue to be studied to better tune the pore network of these materials and explore novel applications.

A great advantage of conventional activation of organic gels over the carbon precursors is the possibility of coupling the mesopore network of the organic gels with micropores created during carbonization or activation step. Tamon et al. studied the effect of carbonization on the porous properties of aerogels and cryogels proving that both precursors preserved the mesoporous structure after the thermal treatment and that cryogels are more prone to develop micropores. Regarding activation, literature data reveal that the gasification of aerogels with CO_2 allows retaining the mesopore network of the organic gels and increasing the volume of micropores. Moreno-Castilla and co-workers reported the activation of a mesoporous aerogel with CO_2 showing that for a 22% burn-off the increase and widening of the precursor micropore network practically did not change the mesopore structure. In a recent work, Ania and co-workers evaluated the effect of carbonization and activation (physical and chemical) of xerogels which, due to their drying in subcritical conditions, may suffer stronger changes upon thermal treatment. In fact, both carbonization and CO_2 activation lead to the increase of the micropore volume at the expense of a severe decrease in the mesopore volume and shrinkage of the average pore size. However, under controlled experimental conditions (i.e. impregnation methodology and temperature), the chemical activation of the xerogel with KOH or K_2CO_3 allowed to suppress the shrinkage and structural collapse, forming a micropore structure associated with the enlargement and/or preservation of the mesopore network of the pristine xerogel.

Hydrothermal carbonization (HTC) is probably one of the most promising alternatives to conventional carbonization of biomass, as it is clearly shown by the ever-increasing number of publications focused in this process since the beginning of the twentieth century. HTC is inspired in the natural processes of coal formation that take millions years and request temperature and pressure in a nonoxidizing atmosphere to transform biomass into a carbon-rich material. Interestingly, the first research work on HTC was published in 1913 by Friedrich Bergius, a Nobel Laureate in recognition of his studies regarding carbon-water reactions at high pressure and mild temperature to successfully produce H_2 by avoiding CO generation. When performing the experiments, Bergius noticed that when peat was used as carbon source the solid residue obtained had an elemental composition similar to that of coal, leading him to investigate the HTC decomposition of plant-based compounds into coal-like materials. As reported in, with the exception of the works by Berl and Schmidt and van Krevelen et al., the interest in the solid carbon materials obtained from HTC process was forgotten until the work published by Wang et al. in 2001. These authors reported the synthesis of carbon spheres from sugar under hydrothermal conditions (190 °C and self-generated pressure in a high-pressure vessel). HTC is a cost effective and eco-friendly process; since to convert biomass into carbon-rich materials, it uses water as solvent, mild temperatures, self-generated pressure and occurs in few hours with no CO_2 emission. Since 2001, the interest in this carbonization process has been exponentially increasing, and in 2016, around 400 papers mentioning hydrothermal carbonization were published, and the works focusing in this process received more than 11,000 citations. Some advantages of HTC-derived materials, commonly known as hydrochars, over charcoals obtained by conventional carbonization is their high content in oxygenated groups and the possibility of morphological control (e.g. spherical morphology).

Hydrochars have an incipient pore network; thus, conventional activation has been commonly employed to obtain specialized nanoporous carbon materials, mainly regarding surface functionalization, morphology features and ultrahigh surface areas and pore volumes. The use of hydrochars as activated carbon precursors was first reported by Zhao et al. aiming to prepare a nitrogen-containing porous carbon material to be used as supercapacitor. The authors performed HTC of D-glucosamine (carbon source) and the hydrochar containing 6.7% of nitrogen was further activated with KOH allowing to obtain a nitrogen-doped porous carbon with a BET area of 600 $m^2\ g^{-1}$.

The great majority of studies concerning hydrochar activation used KOH as activating agent, mainly because, in general, this oxidizing agent allows obtaining the most developed microporosities reaching BET surface areas higher than 3000 $m^2\ g^{-1}$. It is also interesting to notice that there are studies reporting the use of hydrochars for the synthesis of micro + mesoporous materials via chemical activation with zinc chloride or hydroxides, with the mesopore volume corresponding to at least 50% of the total pore volume, i.e. textural characteristics that are not easy to obtain by the most conventional carbonization + activation process. A recent work reports the preparation of an N-doped nanoporous carbon with similar volumes of micro and mesopores and a BET area near 1000 $m^2\ g^{-1}$ by using HTC of bamboo shoots in the presence of water and 0.5 cm^3 of H_2SO_4 followed by the carbonization of the obtained hydrochar, so in the absence of any activating agent. The work by Fuertes and Sevilla enables the synthesis of superactivated porous carbons with 50% of mesopores by adding melamine during the HTC process, thus the final materials also present 1.3–1.7% of nitrogen. Besides opening the possibility of preparing mesoporous carbon materials, the use of hydrochars as activated carbon precursors also allows better control of the micropore size distribution.

Table: Appropriate precursors, kinetic of activation, and type of porosity usually obtained for the most common activating agents.

Type of activation	Activating agent	Appropriate precursors	Porosity	General trend of experimental conditions on pore size distribution (PSD)
Physical	CO_2	Coals and, in less extend, hard lignocellulosic materials	Micro	- High activation degree leads to high volume of micropores with similar pore size distribution (PSD)
	Steam	Coals and, in less extend, hard lignocellulosic materials	Micro (+ meso)	- Widening of PSD with increase in activation degree- Micro + mesopore networks are obtained at high activation degree (yield 20%) and/or high activation temperatures
Chemical	$ZnCl_2$	Lignocellulosic materials (high volatile and oxygen content)	Micro + meso	- Uniform microPSD that widens to the border of micro/mesopore with the increase of Zn/precursor ratio
	H_3PO_4	Lignocellulosic materials (high volatile and oxygen content)	Micro + meso	- PSD mainly in the border of micro/mesopore and dependent on heat treatment temperature (<450 °C)
	KOH	High rank coals (low volatile and high carbon content)	Micro	- KOH/precursor ratio has more influence on adsorption capacity and PSD than activation temperature - Increase of KOH/precursor ratio widens pores from narrow to large micropores and, in less extend, to small mesopores; also hinders the granular morphology (particle disintegration leads to powders)

When starting from simple carbohydrates, the use of HTC followed by activation may enable preparing carbons with spherical morphology (diameters in the micrometer range) and smooth surfaces. It is also possible to obtain a nanoporous carbon material with spherical shape, but with blackberry morphology, by adding acrylic acid to starch during the HTC step.

The use of water as solvent constitutes a major advantage of HTC over convention carbonization that requests dry biomass. In fact, HTC makes feasible the use of wet biomass as is the case of algae or sewage sludge that after being carbonized can be activated to obtain a carbon material with well-developed pore structure and also heteroatom-rich surface (i.e. nitrogen and sulfur). HTC is, in fact, a valuable process when envisaging heteroatom-doped carbon materials, since if precursors with nitrogen and/ or sulfur—e.g. glucosamine, algae, sewage sludge, tobacco rods or bamboo shoots— were selected this methodology allows retaining higher percentages of heteroatoms compared to conventional carbonization, and thus assures relevant amounts of these atoms after the activation step.

Following a somewhat different approach, White et al. reported the preparation of nitrogen-doped carbogel materials by reacting glucose and ovalbumin (secondary biomass precursor) in HTC conditions. The novelty was the saturation of the monolithic hydrochar with ethanol followed by extraction under supercritical conditions, resembling the last step of the synthesis of organic aerogels. The final material was obtained after carbonization (350–900 °C). When using 550 °C, a material with an interconnected 3D pore system, BET area of 476 $m^2 g^{-1}$ and 5–7% (wt.) nitrogen content was obtained.

Very recently, Sevilla et al. proposed an alternative to KOH activation of biomass-derived hydrochars by using potassium oxalate ($K_2C_2O_4$) and melamine to obtain N-doped superactivated carbons with high yields (40–46%). The materials attained BET areas near 3000 $m^2 g^{-1}$, presenting mainly micropores (70%) and nitrogen content between 0.5 and 0.9% irrespectively the amount of melamine used. Although the mixture of potassium oxalate/melamine is presented as an alternative to the corrosive KOH, it is important to highlight that this synthesis route is restricted for lab-scale (under restricted security conditions) since the KCN present in the solid will generate dangerous toxic vapors of HCN during the washing with HCl.

Acid-mediated carbonization (AMC) is so far the less studied alternative method to obtain a carbon-rich material. However, considering that acid catalysis is a common practice to extract sugars from lignocellulosic biomass and that, sugars can be successfully used as activated carbon precursors (i.e. via HTC followed by activation), this process must also be explored. This methodology allows processing largely available and hopefully low cost biomass residues. Moreover, AMC occurs at atmospheric pressure and, in some cases, at low temperature (≈ 100 °C) allowing processing carbon precursors that do not gather the adequate properties to be carbonized through the conventional process or by HTC. In fact, AMC is able to maximize the availability of cellulose and hemicellulose in carbon precursors with high inorganic content (e.g. rice husk) and is also efficient for producing carbon-rich materials from liquid carbon precursors, as is the case of glycerol. Depending on the acid catalyst used (e.g. H_2SO_4 and/or H_3PO_4), the obtained oxygen-rich chars also contain relevant amounts of sulfur or phosphorus groups, thus enhancing the reactivity for subsequent activation and allowing to join a well-developed pore network with heteroatom doping in the final activated carbon.

Wang et al. evaluated sulfuric acid hydrolysis of rice husk followed by the dehydration, polymerization and carbonization of the sugars to yield chars, which can be named as acidchars. These materials were activated with H_3PO_4 and KOH to produce nanoporous carbons with BET areas higher than 2400 $m^2 g^{-1}$, and in the case of phosphoric acid activation, materials with micro + mesopore networks. In 2010, the authors reported the effect of H_2SO_4 concentration during carbonization step, as well as the reaction temperature and time onto the yield of the obtained acidchar that attained 32% (wt.) for the reaction at 95 °C during 10 h when 72% H_2SO_4 solution was used in both hydrolysis and carbonization. Regarding the effect of the H_2SO_4 concentration (42–72%) during acid carbonization onto the textural properties of the activated carbon obtained by H_3PO_4 activation, the authors concluded that although acidchars prepared under distinct acid concentrations present similar elemental analysis, the BET area and pore volumes increase as the concentration of the H_2SO_4 decreases. These findings were rationalized considering that high H_2SO_4 concentrations promote the aggregation of carbon nanoparticles during carbonization, limiting the uniform H_3PO_4 impregnation of the particles and consequently the development of the pore network in the inner particle.

In a recent publication, Cui and Atkinson systematically investigated liquid glycerol AMC using various acid catalysts (i.e. H_2SO_4, H_3PO_4, HCl and CH_3COOH) aiming to study the effect of the acid carbonization conditions onto the textural properties of glycerol-derived nanoporous carbon materials obtained by subsequent physical activation with steam and CO_2. The AMC of glycerol was made under nitrogen between 400 and 800 °C, and the highest carbonization yields were obtained at 400 °C for 10:3 volumetric mixtures glycerol:acid (30% yield for H_2SO_4 and 50% yield for H_3PO_4). Upon activation, materials with BET area values ranging between 990 and 2470 $m^2\ g^{-1}$ and tailored porosity were obtained. The H_3PO_4-char originated micro + mesoporous carbon materials with the volume of mesopores being more than 50% of the total porosity regardless the physical activating agent and the amount of acid during the AMC, while the steam activation of the H_2SO_4-char lead to materials with 40–44% mesopore volume, and the CO_2 activation only 22–25% mesopore volume. The elemental analysis of the nanoporous carbon materials obtained by H_2SO_4 carbonization revealed the presence of 0.35–0.71% of sulfur and the materials carbonized with H_3PO_4 and activated with CO_2 attained even higher heteroatom doping with phosphorus content between 2.04 and 4.34%.

Other Strategies to Synthetize Nanoporous Carbons

Variations of Hydrothermal Carbonization (HTC) Process

In parallel with the studies centered in the activation of hydrochars, in the last few years, the scientific community also started to explore new synthesis routes to obtain a porous structure during the HTC step, being prepared porous carbons with BET surface areas up to 700$m^2\ g^{-1}$. The methodologies proposed avoid the need of further thermal or chemical activation, may enable the synthesis of heteroatom-doped solids and can also allow the synthesis of hierarchical materials.

Fechler et al. reported the synthesis of porous carbon materials with BET areas between 425 and 672 $m^2\ g^{-1}$ through HTC (180 °C overnight) of glucose mixed with several eutectic salt mixtures (i.e. $LiCl/ZnCl_2$, $NaCl/ZnCl_2$, and $KCl/ZnCl_2$) in the presence of a small amount of water. The authors proved that both the amount of water added and the salt composition are determinant for the successful synthesis of materials, which are formed by very small particles aggregation, identical to aerogels. The use of the eutectic mixture $LiCl/ZnCl_2$ originated the carbon materials with the highest surface area. When 2-pyrrol-carboxyaldehyde was added as co-reagent of the mixture glucose and $ZnCl_2$, a material with 3% of nitrogen and BET area of 576 $m^2\ g^{-1}$ was obtained. Fellinger et al. used glucose as carbon source and borax ($Na_2B_4O_7$) as both catalyst and structure-directing agent to prepare hierarchical structured carbogels under HTC conditions. Further carbonization under nitrogen at 550 or 1000 °C allowed to obtain a carbon material with a BET area of 614 $m^2\ g^{-1}$ and 70% of mesopore volume.

Carbonization of Organic Salts

In 2010, Zhou and co-workers developed mesoporous carbons by the carbonization of organic salts (magnesium and barium citrates) evaluating the influence of the temperature of the thermal treatment (600–800 °C). In the case of magnesium salt, BET areas up to 2322 $m^2\ g^{-1}$ were attained, and the increase of temperature resulted in a progressive increase of the mesopore volume percentage (from 50 to 100%). The barium citrate-derived materials are mainly mesoporous (>90%) and, independently of the temperature, pores between 10 and 20 nm are obtained. Calcium citrate was tested by other authors who reported the paramount importance of temperature in the textural properties of the mesoporous carbons.

Atkinson and Rood proposed the use of dichloroacetates of alkaline metals as carbon precursors for the synthesis of nanoporous carbons, being the pore networks, once again, dependent on the cation. The pore structure is produced by fast pyrolysis (15 s to 7 min) under nitrogen flow at temperatures between 300 and 1100 °C, and materials attained BET areas of 740 $m^2\ g^{-1}$. This methodology allowed the synthesis of microporous materials when lithium salt was used and micro + mesoporous solids for the other two metals. In the same research line, Xu and co-workers reported the pyrolysis of EDTA salts to obtain nitrogen-doped porous carbons with BET areas reaching 1800 $m^2\ g^{-1}$ and porosity characteristics dependent on the thermal treatment temperature (higher the temperature, higher the mesopore volume).

The protocol of organic salts carbonization was extended to gluconates and alginates along with citrates to understand the mechanism of the porosity development. The results shown that the textural properties of the porous carbon materials obtained by this methodology are heavily dependent on the type of the cation in the organic salt: while potassium salts originate essentially microporous solids, for sodium and calcium the amount of mesopores is also relevant. In the case of calcium citrate-derived material, the development of the mesoporosity was attributed to the formation of CaO nanoparticles, which act as endotemplates during the carbonization. It was also shown that the nature of the organic salt has a great impact on the morphology, with sodium gluconate leading to the formation of large carbon nanosheets, while sodium citrate originates sponge-like particles. The synthesis of nitrogen-doped porous carbons was also explored by mixing potassium gluconate with melamine, which allowed to obtain a microporous material gathering 22.9% of nitrogen with 660 $m^2 g^{-1}$ of BET area. The endotemplate approach on the carbonization of organic salts was further explored with iron, calcium and zinc citrates. The carbonization of these nonalkali organic salts produces mesoporous materials with BET areas between 950 and 1610 $m^2 g^{-1}$ and distinct pore size distributions: monomodal distribution centered at 11 nm for calcium citrate, bimodal distribution centered at 9 and 20 nm for iron citrate, and bimodal distribution centered at 3 and 10 nm for iron citrate. These carbons can be post-functionalized by heat treatment in the presence of melamine to obtain materials gathering high BET area and mesopore volume with high nitrogen content (8–9%).

Ionothermal Approaches

The synthesis of porous carbon materials via ionothermal concept derives from the methodologies proposed in 2004 by Morris and co-workers to prepare zeolite analogues (i.e. inorganic materials) using ionic liquids or eutectic mixtures as both solvents and inorganic structure directing agents (templates). The ionothermal approach is analogous to hydrothermal and solvothermal processes, where the solvents are predominantly molecular (water or nonaqueous solvents, respectively) but, as the name means, it refers to processes occurring in ionic solvents (e.g. ionic liquids or eutectic mixtures). The low vapor pressure of the ionic solvents is a great advantage of ionothermal reaction over the hydrothermal or solvothermal concepts, since it allows avoiding the safety concerns connected with the high pressures required to prevent molecular solvent evaporation.

The preparation of carbon materials following the ionothermal principles is reported as both ionothermal and molten salt synthesis processes. Nowadays, there is still not a generally accepted terminology for these processes what turns difficult to understand the classifications and underlying procedures followed. In fact, the use of ionothermal/ molten salt process can be linked either to the preparation of carbon materials with incipient porosity obtained at temperatures ≈ 200 °C or to the preparation of porous carbons by a two-step process including the ITC followed by a thermal treatment of the ionothermal derived carbon at high temperatures (attaining 1000 °C or more). Ionothermal/molten salt process is also considered in the case where the mixture of the carbon precursor and the ionic solvent is directly thermally treated at high temperature.

The first studies reporting the preparation of porous carbon materials through ionothermal process used ionic liquids as carbon precursors. In 2009, Lee et al. synthesized N-doped materials (2–3%) with BET area values between 640 and 780 $m^2 g^{-1}$, other approach of the same research group gathered an ionic liquid with simple carbohydrates allowing to obtain carbon materials with a highly developed mesopore network after treatment at only 200 °C during 20 h in a nonpressurized chamber. Xie et al. reported the synthesis of magnetic hierarchical porous carbons by using several carbohydrates and an iron containing ionic liquid.

In the subsequent research in ionothermal approaches, the introduction of salts as co-reagents became common and generally accepted as a hard template route. Zinc chloride is by far the most frequently reported salt, used alone or in mixtures (eutectic or not) with other salts. This strategy allowed to synthesize porous carbons with ultrahigh surface area (easily around 2000 m^2 g^{-1}), hierarchical structure and high heteroatom doping (i.e. nitrogen (>5%) and sulfur). All these features are of fundamental importance for boosting the application of these materials in energy storage processes. While initially nitrogen-containing ionic liquids were used as carbon precursors, along the years, the number of studies exploring carbohydrates, or even biomass, as carbon source has increased.

More elaborated synthesis schemes have been reported, where organic solvents were added dropwise to molten $ZnCl_2$ at 550 °C. By changing the solvent, the authors were able to obtain nitrogen (14%) or sulfur (13%) doped porous carbons with different morphologies (i.e. aerogel, nanosheets or hyperbranch). There are also reports on ITC of biomass with nitrogen-containing ionic liquid followed by conventional KOH activation to yield nitrogen doping up to 1.59% and apparent surface areas of 2838 m^2 g^{-1}. The ionothermal approach is actually a powerful route to synthesize porous carbon materials with valuable graphene-like (2D) structures.

It is also possible to found reports on highly mesoporous carbon materials obtained by $ZnCl_2$-mediated ionothermal/molten salt synthesis. Although these routes are presented as novel synthesis procedures, the high mesopore volumes reported are most probably the result of the complex $ZnCl_2$ activation mechanism when very high amount of $ZnCl_2$ is added. As it was the behavior of chemical activating agents as templates during carbonization cannot be disregarded, and for the particular case on $ZnCl_{2,}$ the results obtained by Molina-Sabio and Rodríguez-Reinoso allowed the authors to conclude that this chemical acts as template for the creation of porosity. So, although new materials are being produced under "novel" synthesis procedures with appealing names, in some cases, the experimental route and underlying mechanism for pore creation seems to be the one accepted for conventional chemical activation.

Zeolite

Zeolites are microporous, aluminosilicate minerals commonly used as commercial adsorbents and catalysts. The term zeolite was originally coined in 1756 by Swedish mineralogist Axel Fredrik Cronstedt, who observed that rapidly heating the material, believed to have been stilbite, produced large amounts of steam from water that had been adsorbed by the material.

Zeolites occur naturally but are also produced industrially on a large scale. As of December 2018, 245 unique zeolite frameworks have been identified, and over 40 naturally occurring zeolite frameworks are known. Every new zeolite structure that is obtained is examined by the International Zeolite Association Structure Commission and receives a three letter designation.

Natural zeolite with a U.S. penny for scale.

Properties and Occurrence

A form of thomsonite (one of the rarest zeolites).

Zeolites have a porous structure that can accommodate a wide variety of cations, such as Na^+, K^+, Ca^{2+}, Mg^{2+} and others. These positive ions are rather loosely held and can readily be exchanged for others in a contact solution. Some of the more common mineral zeolites are analcime, chabazite, clinoptilolite, heulandite, natrolite, phillipsite, and stilbite. An example of the mineral formula of a zeolite is: $Na_2Al_2Si_3O_{10}\cdot 2H_2O$, the formula for natrolite. These cation exchanged zeolites possess different acidity and catalyse several acid catalysis.

Natural zeolites form where volcanic rocks and ash layers react with alkaline groundwater. Zeolites also crystallize in post-depositional environments over periods ranging from thousands to millions of years in shallow marine basins. Naturally occurring zeolites are rarely pure and are contaminated to varying degrees by other minerals, metals, quartz, or other zeolites. For this reason, naturally occurring zeolites are excluded from many important commercial applications where uniformity and purity are essential.

Zeolites are the aluminosilicate(AlO_4^{5-} and SiO_4^{4-}) members of the family of microporous solids known as "molecular sieves", and mainly consist of Si, Al, O, and metals including Ti, Sn, Zn, and so on. The term molecular sieve refers to a particular property of these materials, i.e., the ability to selectively sort molecules based primarily on a size exclusion process. This is due to a very regular pore structure of molecular dimensions. The maximum size of the molecular or ionic species that can enter the pores of a zeolite is controlled by the dimensions of the channels. These are conventionally defined by the ring size of the aperture, where, for example, the term "8-ring" refers to a closed loop that is built from eight tetrahedrally coordinated silicon (or aluminium) atoms and 8 oxygen atoms. These rings are not always perfectly symmetrical due to a variety of causes, including strain induced by the bonding between units that are needed to produce the overall structure, or coordination of some of the oxygen atoms of the rings to cations within the structure. Therefore, the pores in many zeolites are not cylindrical.

Zeolites transform to other minerals under weathering, hydrothermal alteration or metamorphic conditions. Some examples:

- The sequence of silica-rich volcanic rocks commonly progresses from:
    - Clay → quartz → mordenite–heulandite → epistilbite → stilbite → thomsonite–mesolite-scolecite → chabazite → calcite.
- The sequence of silica-poor volcanic rocks commonly progresses from:
    - Cowlesite → levyne–offretite → analcime → thomsonite–mesolite-scolecite → chabazite → calcite.

Production

Industrially important zeolites are produced synthetically. Typical procedures entail heating aqueous solutions of alumina and silica with sodium hydroxide. Equivalent reagents include sodium aluminate and sodium silicate. Further variations include changes in the cations to include quaternary ammonium cations.

Synthetic zeolites hold some key advantages over their natural analogues. The synthetic materials are manufactured in a uniform, phase-pure state. It is also possible to produce zeolite structures that do not appear in nature. Zeolite A is a well-known example. Since the principal raw materials used to manufacture zeolites are silica and alumina, which are among the most abundant mineral components on earth, the potential to supply zeolites is virtually unlimited.

Natural Occurrence

Conventional open-pit mining techniques are used to mine natural zeolites. The overburden is removed to allow access to the ore. The ore may be blasted or stripped for processing by using tractors equipped with ripper blades and front-end loaders. In processing, the ore is crushed, dried, and milled. The milled ore may be air-classified as to particle size and shipped in bags or bulk. The crushed product may be screened to remove fine material when a granular product is required, and some pelletized products are produced from fine material.

Natrolite.

As of 2016 the world's annual production of natural zeolite approximates 3 million tonnes. Major producers in 2010 included China (2 million tonnes), South Korea (210,000 t), Japan (150,000 t), Jordan (140,000 t), Turkey (100,000 t) Slovakia (85,000 t) and the United States (59,000 t). The ready availability of zeolite-rich rock at low cost and the shortage of competing minerals and rocks are probably the most important factors for its large-scale use. According to the United States Geological Survey, it is likely that a significant percentage of the material sold as zeolites in some countries is ground or sawn volcanic tuff that contains only a small amount of zeolites. Some examples of such usage include dimension stone (as an altered volcanic tuff), lightweight aggregate, pozzolanic cement, and soil conditioners.

Artificial Synthesis

Synthetic zeolite.

There are over 200 synthetic zeolites that have been synthesized by a process of slow crystallization of a silica-alumina gel in the presence of alkalis and organic templates. Many more such structures could theoretically be made. In addition to variations in structures, zeolites can also be made with a variety of other atoms in them to make them chemically interesting and active. Some examples of the so-called heteroatoms that have been incorporated include germanium, iron, gallium, boron, zinc, tin, and titanium. One of the important processes used to carry out zeolite synthesis is sol-gel processing. The product properties depend on reaction mixture composition, pH of the system, operating temperature, pre-reaction 'seeding' time, reaction time as well as the templates used. In sol-gel process, other elements (metals, metal oxides) can be easily incorporated. The silicalite sol formed by the hydrothermal method is very stable. The ease of scaling up this process makes it a favorite route for zeolite synthesis.

The Zeolite Conundrum

Computer calculations have predicted that millions of hypothetical zeolite structures are possible. However, only 232 of these structures have been discovered and synthesized so far, so many zeolite scientists question why only this small fraction of possibilities are being observed. This problem is often referred to as "the bottleneck problem". Currently there are a number of theories attempting to explain the reasoning behind this question:

- Zeolite synthesis research has primarily been concentrating on hydrothermal methods; however, new zeolites may be synthesized using alternative methods. Synthesis methods that have started to gain use include: Microwave-assisted, post-synthetic modification, steam.
- Geometric computer simulations have shown that the discovered zeolite frameworks possess a behaviour known as "the flexibility window". This shows that there is a range in which the zeolite structure is "flexible" and can be compressed but retain the framework structure. It is suggested that if a framework does not possess this property then it cannot be feasibly synthesised.
- As zeolites are metastable, certain frameworks may be inaccessible as nucleation cannot occur because more stable and energetically favourable zeolites will form. Post-synthetic modification has been used to combat this issue with the ADOR method, whereby frameworks can be cut apart into layers and bonded back together by either removing silica bonds or including them.

Mesoporous Silicate

Mesoporous silicates are silicates with a special morphology. Porous inorganic solids have found great utility as catalysts and sorption media because of their large internal surface area, i.e. the presence of voids of controllable dimensions at the atomic, molecular, and nanometer scales. With increasing environmental concerns worldwide, nanoporous materials have become more important and useful for the separation of polluting species and the recovery of useful ones. In recent years there has been great progress in applying environmentally friendly zeolites in heterogeneous reaction catalysis. The reason for their success is related to their specific features in converting molecules having kinetic diameter below 1 nm, but they become inadequate when reactants with sizes above the dimensions of the pores have to be processed. Research efforts to synthesize zeolites with larger pore diameter, high structural stability and catalytic activity have not given the expected results yet.

Characteristics

The discovery of a new family of mesoporous molecular sieves in the early 1990s by Kuroda et al., known as KSW-1 and FSM-16, and by ExxonMobil, called M41S, opened new possibilities to prepare catalysts for reactions of relatively large molecules. The silicate wall of the pores is amorphous. Mesoporous silicates, such as MCM-41 and SBA-15 (the most common mesoporous silicates), are porous silicates with huge surface areas (normally ≥1000 m^2/g), large pore sizes (2 nm ≤ size ≤ 20 nm) and ordered arrays of cylindrical mesopores with very regular pore morphology. The large surface areas of these solids increase the probability that a reactant molecule in solution will come into contact with the catalyst surface and react. The large pore size and ordered pore morphology allow one to be sure that the reactant molecules are small enough to diffuse into the pores.

Synthesis of MSNs

Stober was the pioneer in developing a system of chemical reactions for the synthesis of spherical monodisperse micron size silica particles. From then on, the method is known as Stober synthesis. Many modifications have constantly been made to the Stober's synthesis to yield monodisperse, ordered, nanosized silica particles. The synthesis of MSNs can be accomplished in basic, acidic and neutral conditions. Manipulating the reaction parameters resulted in particles with different shapes and sizes. The Stober's method of synthesis was first modified by Grun et al. where they introduced a cationic surfactant as a template to yield a spherical rather than a hexagonal MCM-41 structure. They were successful in generating spherical MCM-41 with similar properties as that generated by other methods. Constant research has led to a lot of variations in the synthesis conditions and methods to yield stable, monodisperse MSNs.

For MSNs to be an ideal carrier for drug delivery the particle size needs to be uniform; pore volume has to be large to enhance loading capacity. These parameters can be controlled during the synthesis by varying the pH of the reaction mixture, temperature, concentration of surfactant and silica source. The synthesis of MSNs occurs by liquid crystal template mechanism wherein hydrolysis and condensation of silica on the surface of surfactant micelles takes place. The liquid silica (tetraethyl orthosilicate) transforms to solid silica.

Mechanism of Formation of MSNs

A thorough understanding of the mechanism of formation of MSNs is essential to obtain particles with desired properties for drug delivery. The early reports on the mechanism suggested that the silica network gets built throughout the liquid–crystalline phases of non-ionic surfactants. This is particularly true for materials prepared from a dilute solution of surfactants as no evidence of regular mesostructured materials was observed. either the hydrolysed silica gets adsorbed around the micelles or in the case of SBA-15, the surfactant and the silica interact at the initial stage and form a core shell-like structure. The mechanism for the formation of MCM-41 is represented in figure. Efforts have been on since then by research groups to unveil the exact mechanism behind the formation of MSNs.

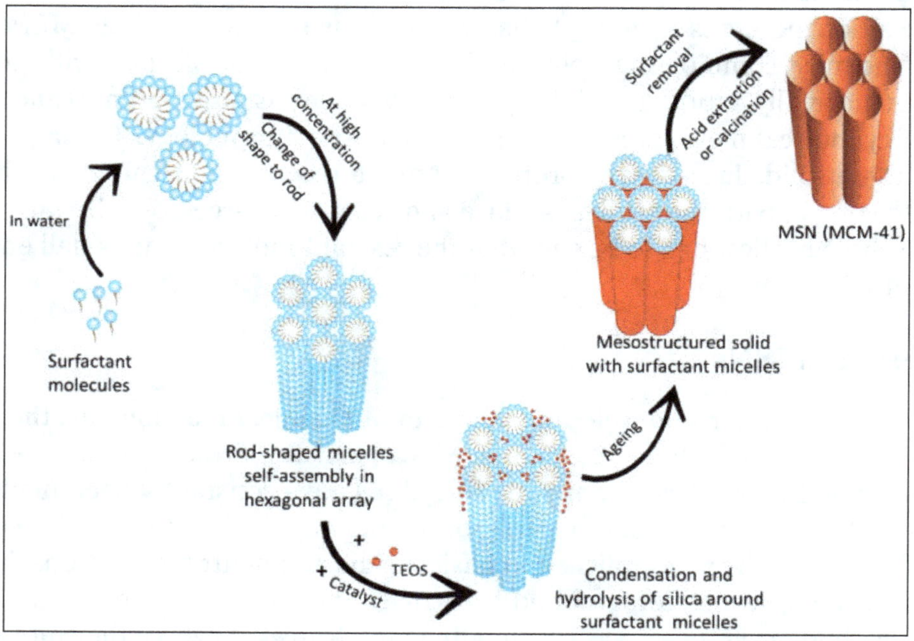

Mechanism of formation of Mobil Crystalline Materials No.41 (MCM-41).

The in-situ usage of time-resolved small-angle neutron scattering (SANS) has been used to study the formation of MSNs. Using this method, they were able to predict the changes happening concurrently with the formation process. It was observed that during the early hydrolysis (~40 s) of silica source tetramethyl orthosilicate (TMOS), the silicate ions tend to adsorb around the surfactant micelles during the growth phase. As the charge around the surfactant reduces due to the initial hydrolysis and the condensation of the silica precursor, the intermicellar repulsion reduces, allowing the further formation of small aggregates of silica. After ~400 s, the reaction mixture contained sufficiently discrete hexagonally ordered mesopores of silica which was confirmed by transmission electron microscopy (TEM) studies.

Another mechanism named 'swelling-shrinking mechanism' was proposed for the formation of MSNs utilizing the technique of time-resolved synchrotron small-angle X-ray scattering (SAXS). This mechanism holds well when tetraethyl orthosilicate (TEOS) alone is used as the precursor in the absence of any other solvent like ethanol. TEOS being oil-like monomer showed phase separation under static condition, whereas, under vigorous stirring, an emulsion-like system was obtained. Initially, cetyltrimethylammonium bromide (CTAB) forms ellipsoidal micelles with an inner core consisting of the hydrophobic tail. When TEOS is added, it gets solubilized in the hydrophobic core, thus enlarging the micelles and resulting in the transformation of micelle shape from ellipsoidal to spherical. On hydrolysis of TEOS, the monomers become hydrophilic and are released into the aqueous surroundings. The negatively charged hydrolysed monomers of TEOS get adsorbed onto positively charged CTAB micelles via electrostatic attraction. On complete consumption of the TEOS within the hydrophobic core, the micelles shrink and become smaller in size. As this process of hydrolysis and condensation occurs simultaneously, the micelles shrink continuously until all the TEOS gets hydrolysed and form silica shell around the micelles. The neighbouring micelles aggregate, resulting in particle growth forming a mesoporous structure.

Approaches for the Synthesis of MSNs

Majority of the MSNs are fabricated by modified Stober's method otherwise popularly known as a sol-gel process. Sol-gel chemistry is a widely explored process for the synthesis of many inorganic materials. It involves the hydrolysis and condensation of the alkoxide monomers into a colloidal solution (sol), which acts as a precursor to form an ordered network (gel) of polymer or discrete particles. A typical sol-gel process takes place in the presence of an acid or a base catalyst. Depending on the reaction conditions and the molar ratio of Si/H_2O, the alkoxide group gets hydrolysed. The rate of hydrolysis proceeds faster in basic conditions compared to acidic. Condensation succeeds the hydrolysis step and the effective condensation depends on the hydrolysis step. Multiple condensation results in a chain-like structure in the sol and network-like structure in gel form. The schematic representation of the reaction is shown.

$$\equiv Si\text{-}OH + OH\text{-}Si\equiv \leftrightarrow \equiv Si\text{-}O\text{-}Si\equiv + HOH$$
$$\equiv Si\text{-}OH + RO\text{-}Si\equiv \leftrightarrow \equiv Si\text{-}O\text{-}Si\equiv + ROH$$
$$\cdots\cdots \longrightarrow \text{polysilicates}$$

"R' may be methyl, ethyl, propyl, butyl etc.

To yield particles of the desired size and to enhance the properties of MSNs, the sol-gel process has been modified.

A simple quenching approach was adopted by Mann et al. to synthesize small sized and ordered MSNs. The reaction was quenched after 40 s by the addition of an excess of water followed by neutralization to pH 7 with dilute hydrochloric acid after a delay time ranging from 60 s to 220 s. The rest of the experimental condition and reagents were maintained the same as in the routine procedure. It was observed that dilution reduces coalescence, and neutralization reduces the rate of silica condensation. The results showed that greater the delay in the neutralization, larger the particle size. Similar results were observed by Moller group. The colloidal suspension obtained was found to be stable for long periods of time. However, the MSNs produced by this technique were found to be less ordered in a structure which could be due to scale-up issues during dilution.

Evaporation-induced self-assembly (EISA) is another approach for the synthesis of MSNs. In this technique, all the reactants undergo concentration changes during evaporation throughout the process. This results in the organization of a liquid-crystal like template of the silica precursor. In this method, the required concentration of precursor formulations was prepared in ethanol/water solvent with a surfactant. This was converted to monodisperse droplets via injection into an aerosol generator. The droplet size can be controlled by altering its orifice. The alcohol evaporation during drying induces micelle formation and the co-assembly of silica-surfactant into liquid-crystal mesophases. Fontecave et al. modified the EISA method by incorporating different amphiphilic drugs (stearoyl choline, sophorolipid and glucosyl-resveratrol) which behave both as a structure directing agent as well as an active cargo. Sol compositions containing TEOS, drug, water, ethanol and HCl were prepared and injected into an aerosol apparatus. A spray dryer with sufficient air flow and pressure was provided to convert the droplets into solid particles. Characterization of the particles for its pore size and structure using TEM revealed similar results to those with CTAB as surfactant template. Even though the mesostructure formation was not that good, the drug loading was found to be high in all the three cases with complete elimination of burst release. This method would be an ideal strategy for hydrophilic drugs functionalized with hydrophobic tails. Due to the absence of the surfactant template, intrinsic toxicity can be reduced.

A recent widely used modification to the synthesis of parent MSNs is encapsulating drugs in hollow mesoporous silica nanoparticles (HMSNs). Their large hollow cavity inside each MSN has garnered tremendous attention. These hollow cavities are capable of holding a high amount of drug compared to its non-hollow counterparts. This unique property of HMSNs makes it widely useful in cancer therapy and imaging. Shi et al. were one of the first groups to report the synthesis of HMSNs. Preliminary studies to ascertain its enhanced loading capacity was performed using ibuprofen as the model drug wherein, HMSNs showed an enhanced drug loading of 744.5 mg/g, as compared to 358.6 mg/g of MCM-41. The frequently used method for the synthesis of HMSNs includes the 'core-templating method'. In this approach, many soft/hard templates are used to form the core followed by coating with desired substance at different concentrations to obtain a shell around the substrate with a desired thickness. Subsequently, the core template can be eliminated by calcination or treatment with a suitable solvent leaving behind the shell with a hollow core. She et al. synthesized HMSNs using eudragit S-100 and Triton X-100 as the core-shell template. HMSNs using SiO_2@CTAB-SiO_2 nanoparticles as templates were synthesized by selective etching process. Ghasemi et al. synthesized HMSNs using poly tert-butyl acrylate (PtBA) nanospheres as core-forming hard templates in the presence of CTAB as soft templates.

Many hard/soft templates like vesicles, polymeric micelles, gold or silica nanoparticles were used to construct templates. Lin and collaborators proposed a new technique for the synthesis of MSNs using water-in-oil microemulsion as a template. The advantages of this method were the uniformly sized particles obtained compared to other methods. Also, the microemulsion is said to be thermodynamically stable.

As the commonly used sol-gel process is a laborious, time consuming process, many different fast methods were used for the synthesis of MSNs. Ling and Su proposed a low-cost electrochemistry assisted approach to synthesize MSNs in large-scale. The formation of MSNs was accomplished by the production of hydroxide at the stainless steel substrate/solution interface which resulted in the self-assembly of surfactant micelles and the polycondensation of silica precursors. Microwave assisted technique for the synthesis of MSNs could be another low-cost approach for the synthesis of MSNs. Various reports show that MSNs with ordered pore size and arrangement could be rapidly synthesized by this method. Another rapid, cost effective method for the fabrication of MSNs is sonochemcial synthesis. The use of photoacoustic cavitations during the process was found to generate ordered MSNs, giving scope for the fine tuning of the process in a shorter duration of time.

Raw Materials used and Factors affecting the Characteristics of MSNs

The three main elements that form the heart of MSN includes a silica precursor (tetraethyl orthosilicate-TEOS, tetramethyl orthosilicate-TMOS, tetramethoxyvinylsilane-TMVS, sodium meta-silicate and tetrakis(2-hydroxyethyl) orthosilicate- THEOS), a surfactant (non-ionic or cationic surfactant) as a structure directing agent (SDA) and a catalyst. Other additives like cosolvents, compounds to prevent aggregation may also be incorporated based on the requirements. In order to ensure the scale-up of MSNs at reasonable cost, natural perlite materials like pumice rock, rice husk, and renewable biomass could also be explored for the synthesis of MSNs. The common chemical constituents explored so far are listed in table.

Table: List of commonly used chemicals in the synthesis of MSNs.

Chemical Constituents	Function
Cetyltrimethylammonium bromide (CTAB)	Structure directing agent/template
Cetyltrimethylammonium chloride (CTAC)	Structure directing agent/template
Pluronic F123, F127	Surfactant template
Brij-76	Surfactant template
Triton X-100	Surfactant
Tween 20, 40, 60, 80	Surfactant
Tetraethyl orthosilicate (TEOS)	Inorganic silica source
Tetramethoxy silane (TMOS)	Inorganic silica source
Tetrakis(2-hydroxyethyl) orthosilicate (THEOS)	Inorganic silica source
Trimethoxyvinylsilane (TMVS)	Inorganic silica source

Sodium silicate	Inorganic silica source
Ethanol	Cosolvent to solubilize TEOS
Sodium hydroxide (NaOH)	Base catalyst
Ammonium hydroxide (NH_4OH)	Base catalyst
Triethanolamine (TEA)	Base catalyst, complexing agent and growth inhibitor
Diethanolamine (DEA)	Base catalyst
Disodium hydrogen phosphate-sodium dihydrogen phosphate buffer solution	Reaction medium
Triisopropylbenzene (TIPB)	Pore-expanding agent
Tetrapropoxysilane (TPOS)	Pore- expanding agent
Pluronic polymer P103	Pore-expanding agent

The particle size, pore size and morphology of MSNs can be successfully modulated as required by varying the reaction conditions (relative amounts of alkoxysilane, water, catalyst) and temperature.

Control of Particle Size

Particle size is a very important feature for the biomedical application of MSNs as drug carriers. Hence careful tuning of particle size is essential for effective drug delivery. pH of the reaction medium plays a pivotal role in governing the size of MSNs. The particle size can be effectively controlled by adding suitable additive agents like alcohols, amine, inorganic bases and inorganic salts. These agents alter the hydrolysis and condensation of silica precursor. They accelerate the reaction kinetics thus resulting in particles of smaller size. Moller et al. replaced the often used base catalyst sodium hydroxide (NaOH) and ammonium hydroxide (NH_4OH) with triethanolamine (TEA). In addition to conferring a basic pH, it also acts a complexing agent to obtain discrete nanoparticles. When the molar ratio of TEOS: TEA was changed from 1:1 to 1:4, the largest particle size was observed with the ratio 1:4. Qiao suggested that the initial pH value of the system greatly affects the particle size of MSNs. When they provided Na_2HPO_4-NaH_2PO_4 as a source of OH^- ions, particle size was found to increase with an increase in the initial pH of the solution. El-Toni et al. observed that on increasing the concentration of ammonia beyond a certain level, the agglomeration of silica particles takes place due to the increase in the ionic strength of the reaction medium. Bouchoucha et al. also reported the efficiency of TEA in producing well-dispersed nanoparticles. Use of l-lysine as a base catalyst was found to hinder the growth of silica particles, thus resulting in sub-nanometre size particles. This may be due to the electrostatic interaction between the protonated amine groups of l-lysine and a deprotonated hydroxyl group on silica surface further delaying the condensation process. PEG-silane capping on the surface of silica particles was also found to effectively attenuate the particle growth process by steric stabilization. When PEG-silane was added immediately after addition of TMOS, the particle size was found to be 5 nm but when it was added at a delay of 50 min after addition of TMOS, the particle diameter increased to >13 nm. A change from monodisperse to heterogeneous particle size distribution was observed when the amount of silica precursor TEOS was increased, which may be attributed to the secondary condensation reactions taking place due to the presence of excess silica precursor which starts producing new nuclei amongst the already existing silica particles. A similar trend of results was obtained by Chiang et al. who found that particle size was found to increase with an increase in the amount of TEOS. An increase in the particle size was observed by using different tetraalkoxysilane with different alkoxy groups ($Si(OR)_4$, R = Me, Et, Pr and Bu). Along with this, the addition of alcohols also influenced the particle size of the MSNs. This may be due to the alteration in the hydrolysis rate. Low concentration of CTAB as surfactant yielded a homogenous spherical particle size distribution. On investigating the effect of CTAB on the particle size, it was observed that transformation from discrete spherical to agglomerates was observed due to variation in the hydrolysis and micellization of CTAB. By tuning the concentration of F127 polymer, the particle size could be controlled. An increase of particle size up to 300 nm was reported with an increase in the triblock copolymer Pluronic F127 concentration. Reports state that a balance between the molar composition of various reactants was necessary to obtain MSNs with desired size and quality. The reaction parameters equally influenced the mean particle size of MSNs. On increasing the temperature of the reaction from 30 to 70 °C, an increase in the particle size from 28.91 nm to 113.22 nm was observed. This may be due to the increase in the rate of reaction leading to polycondensation of the silica monomers resulting in a dense silica structure and a larger size.

Control of Pore Size, Pore Volume and Mesostructural Ordering

Depending on the type of surfactant, the pore size of MSNs can be varied. The longer chain length of surfactant results in MSNs with larger pores and those with short chain length gives MSNs with smaller pores. The concentration of TEOS influenced the mesostructural ordering of the particles. The higher amount of TEOS showed a disordered mesostructure whereas lesser amount was not sufficient to form a mesoporous structure. The concentration of surfactant CTAB was also found to have a profound impact on the mesostructural arrangement of the particles. A lower concentration of surfactant fails to form micelles and hence the resulting nanoparticles will be template deficient whereas the too high concentration of CTAB may result in a disordered structure. Hence an optimum balance has to be struck between various reagents used. Moreover, mesostructural ordering takes place in dilute aqueous solutions. Addition of N, N-dimethylhexadecylamine (DMHA) behaves as a pore size mediator and thus helps with efficient control of pore size as per requirements. Pore size was also found to have a profound influence on the release rate of the drugs as observed using ibuprofen. The selection of surfactant species greatly influences the mesostructural ordering of the nanoparticles and the pore size. Effect of the templating agent due to change in the counterions present in cetyltrimethylammonium was studied. Cetyltrimethylammonium chloride (CTAC) as pore generating template produced MSNs with the wormhole-like arrangement.

On changing the counter ion to a much larger tosylate ion (CTATOS), the pore radius was found to increase and the pore morphology changed from wormlike to stellate. Of late, novel strategies have been adopted by scientists to bring about modifications in the properties of the MSNs in order to overcome the drawbacks of the traditional MSNs such as small pore size and poor particle size uniformity. In this regard, Huang et al., synthesized highly monodisperse silica nanoparticles having a large pore size with dendritic morphology using a novel dual templating sol-gel reaction by mixing partially fluorinated short chain anionic fluorocarbon surfactant, Capstone FS-66 and CTAB. An interesting observation was made by them showing that an increase in the amount of Capstone FS-66 incorporated resulted in a change in the morphology. The particle size was larger with a dendritic channel pore structure. As the amount of Capstone is further increased, the morphology transforms into a flower-like large dendritic structure. The same strategy was used by Yu et al., who synthesized dendritic MSNs with particle size of less than 200 nm using imidazolium ionic liquids with different alkyl lengths as cosurfactants and Pluronic F127 as a particle growth inhibitor. They observed that neither the reaction temperature nor the time showed any influence on the particle size of MSNs.

Control of Shape

The shape of the MSNs greatly affects the cellular uptake and biodistribution of MSNs. Hence stringent control of the shape of MSNs is important to regulate its excretion and other effects in vivo. Till date, spherical MSNs have been widely explored for their drug delivery potential. However, the non-spherical MSNs are seldom used. By carefully controlling the reaction conditions, non-spherical materials with rod, ellipsoid, film, platelet, sheet and cube shapes could be generated. The molar concentration of surfactant, water, base catalyst and TEOS was found to have an impact on the morphology of the MSNs. Cai et al. generated MSNs with different shapes like spherical, silica rods and micrometre-sized oblate silica by manipulating the concentration of TEOS, $NaOH/NH_4OH$ and CTAB. By regulating the amount of dodecanol as a soft template and the temperature of synthesis, a broad range of silica particles could be realized varying from sphere to shell, rugby, peanut, hollow and yolk shell-like structures. The six different particles fabricated had controlled size, porosity, interior spaces and shell structure. Their results showed that inclusion of dodecanol as a soft template resulted in particles with different morphologies. This indicates that any changes in the micelle structure at the initial stage leads to a change in the particle morphologies. MSNs of various shapes can be obtained with a wide range of aspect-ratio. Rod-shaped MSNs are widely exploited counterparts of spherical MSNs. These can be obtained by varying the reaction parameters in a typical sol-gel reaction. An increase in the amount of catalyst and addition of co-solvent like heptane, change in temperature, varying the molar composition of reactants yield rod-shaped MSNs. Ellipsoidal shaped MSNs also form a part of drug delivery carrier. Inability to retain its shape due to minimization of surface free energy leading to spherical particles poses a major challenge. These MSNs can be synthesized by the introduction of a co-surfactant, the addition of potassium chloride and ethanol. Platelet-shaped MSNs with high pore accessibility can be synthesized by addition of low amounts of ammonium fluoride and heptane, the presence of non-ionic block copolymer P104, using a ternary surfactant system of cetyltrimethylammonium bromide–sodium dodecyl sulfate–Pluronic.

Apart from the silica precursors, certain organosilanes were incorporated, which performed the dual function of shape transformation and surface functionalization. Morphological variants of MSNs could be synthesized by the co-condensation method of incorporation of organosilanes. The particle morphology depends on the type and amount of the organoalkoxysilane precursors introduced. Various shapes of MSNs such as spheres, rods and hexagonal tubes could be generated using 3-aminopropyltrimethoxysilane (APTMS), N-(2-aminoethyl)-3-aminopropyltrimethoxysilane (AAPTMS), 3-[2-(2-aminoethyl amino) ethylamino] propyltrimethoxysilane (AEPTMS), ureidopropyltrimethoxysilane (UDPTMS), 3-isocyanatopropyltriethoxysilane (ICPTES), 3-cyanopropyltriethoxysilane (CPTES) and allyltrimethoxysilane (ALTMS) as organoalkoxysilanes. The transformation in shape of the MSNs may be attributed to the different types of interaction such as hydrogen bonding, hydrophobic interactions between the organoalkoxysilane and the surfactant template.

Drug Loading and Release of Drugs from MSNs

The unique feature of MSNs which makes it a widely exploited carrier for drug delivery is its high loading capacity due to the large pore volume and surface engineering properties both on the external and internal surface for better drug targeting.

Drug Loading

The drug loading is mainly based on the adsorptive properties of MSNs. Both hydrophilic and hydrophobic cargos can be incorporated into the pores of MSNs. Owing to their large pore volume, MSNs inherently possess greater loading capacity compared to other carriers. Nevertheless, extensive work has been carried out to further enhance the loading of the drugs. Synthesis of HMSNs is one such approach to enhance the loading of MSNs. She et al. attempted to increase the loading of 5-fluorouracil (5-FU) into hollow MSNs by functionalizing the surface silanol groups with different silanes viz, octadecyltrimethoxysilane (OTMS), (3-aminopropyl) triethoxysilane (APTES), 3-cyanopropyltriethoxysilane (CPTES). An improved loading of 28.89% was observed for amine functionalized HMSNs compared to plain HMSNs with 18.34%. This may be via the electrostatic interactions between the negatively charged 5-FU and positively charged amino modified HMSNs. A similar strategy could be used for improving the loading capacity of drugs by electrostatic attractions by varying the type of functionalization. However, a contrasting theory was put forth by Wang et al. whose paper suggested that loading of a drug prior to surface grafting yielded a carrier with high loading as compared to grafting followed by loading of the drug. Compared to MSNs, HMSNs proved to be a better carrier in terms of loading capacity due to their hollow cavities. 3–15 times higher loading of drugs was observed in HMSNs when compared to MSNs. In addition, dual loading of drugs was also achieved using the same carrier. The loading capacity of MSNs could be further enhanced by utilizing polymer gatekeeping for the entrapment of hydrophobic drugs . Consecutive drug loading process which increases the intermolecular interactions can also lead to improved loading of the drugs. An increase in the drug feeding ratio was also found to have a profound influence on the loading capacity of MSNs. The pore volume of MSNs is the major factor which dictates the loading of the drug. Hence pore expansion strategy can be adopted to introduce and hold a large amount of cargo. Pore swelling agents such as alkanes/ethanol, triisopropyl benzene (TIPB), trioctylamine (TOA), decane and N,N-dimethylhexadecylamine (DMHA) aid pore expansion. Table gives a comparison of the drug loading capacity of various MSNs.

Table: Comparison of loading in MSNs.

Carrier	Drug	Loading (wt %)
MCM-41	Ibuprofen	35.9
HMSNs		74.5
MCM-41	Doxorubicin	48.16
HMSNs		112.12
HMSNs	5-fluorouracil	18.54
HMSNs-NH_2		28.89
HMSNs-COOH		20.73
HMSNs-CN		22.54
HMSNs-CH_3		12.13
MCM-$41_{(C12)}$	Captopril	23.6
MCM-$41_{(C16)}$		34
SBA-15		22.6
MCM-41	Erythromycin	29
SBA-15		34
SBA-15 (C8)		13
SBA-15 (C18)		18
MCM-41	Alendronate	14
MCM-41-NH_2		37
SBA-15		8
SBA-15-NH_2		22
MSN-C0	Lysozyme	34
MSN-C10		42

Release of Drugs from MSNs

The release profile of drugs from MSNs mainly depends on its diffusion from the pores which can be tailored by modifying the surface of the MSNs to suit the biological needs.

The decisive factor responsible for controlling the release is the interaction between the surface groups on pores and the drug molecule. It was observed that drug loading followed by surface functionalization with amine groups played a significant role in sustaining the drug release as compared to the systems which were functionalized first and then loaded with the drug. This could be attributed to the loading of the drugs within the pores and the capping with APTES which prevents the drug release. If the surface was first functionalized and then loaded, there are possibilities that the drug will get adsorbed on the surface of MSNs resulting in burst release. The role of APTES concentration in drug release was studied, the results of which revealed that a change in the APTES concentration played a vital role in controlling the drug release from the pores. Aspirin loading and release from the MSNs were studied by post-synthetic grafting as well as co-condensation method. It was observed that co-condensation method showed a greater drug loading compared to the other method. In case of plain MCM-41, weak interaction between aspirin and silanol groups resulted in faster drug release following Fick's diffusion. With amino functionalized MCM-41, the strong interaction between the amine group and aspirin resulted in slow drug release especially for the co-condensed MCM-41. Echoing similar results, the release of ibuprofen from SBA-15 was found to be greatly influenced by the surface modification. In case of amino-functionalized SBA-15 by one-pot synthesis, complete drug release was observed at the end of 10 h whereas the release from that of post-synthetically modified SBA-15, the release of ibuprofen up to 3 days was observed . Table presents the comparison of release rates of different MSNs.

Table: Comparison of release rate of MSNs.

Carrier	Drug	Release Rate
MCM-41 (C12)	Captopril	45 wt % within 2 h, total drug release over 16 h
MCM-41 (C16)		47.47 wt % within 2 h, total drug release >30 h
SBA-15		60 wt % within 0.5 h, total drug release over 16 h
SBA-15	Erythromycin	60% release within 5 h, total drug release within 14 h
SBA-15 (C8)		
SBA-15 (C18)		
SBA-15 unmodified (PS0)	Ibuprofen	Complete release in 10 h
SBA-15-NH$_2$ by post synthesis (PS2)		Initial burst release of 50% in 10 h followed by 100% release in 3 days
SBA-15-NH$_2$ by one pot synthesis (OPS2)		Complete release in 10 h
MSN (grafting-loading approach)	Doxorubicin	40% in 8 h and stagnant release beyond 8 h
MSN (loading-grafting approach)		10% in first 24 h, sustained beyond 160 h

Applications of MSNs in Drug Delivery

MSNs have been widely utilized for a variety of purposes ranging from its use in medicine, as a catalyst in chemical synthesis, adsorbents to adsorb wastes, toxic substances and also as sensors. One of the advantages of MSNs is the ease with which it can be functionalized based on the requirements for a wide variety of applications to control the release of drugs. Figure shows a pictorial representation depicting the versatility of MSN carrier. Vast research is being conducted in utilizing these carriers for drug delivery.

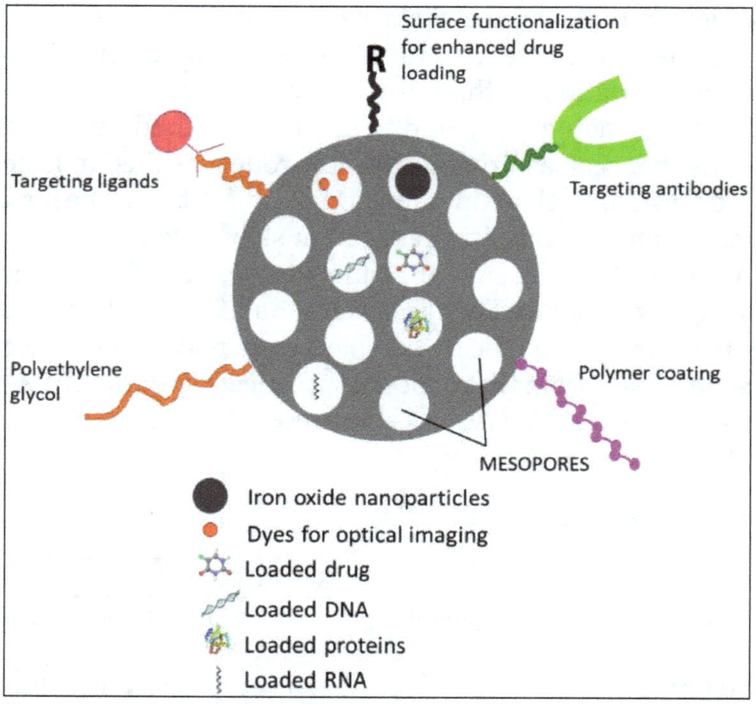

Illustration of versatility of MSN as a carrier in loading variety of drugs.

Targeted Antitumor Therapy using MSNs

The property of surface functionalization of MSNs is widely explored to enhance the site-specific delivery of drugs and avoid side effects. It is a versatile carrier where drugs with different physicochemical properties can be loaded and further functionalized for effective therapy. They can accumulate significantly in the tumour due to the enhanced permeation and retention (EPR) effect. Active targeting aids in maximizing the uptake of actives into the cells. Based on the difference between the normal and tumour cells, suitable receptors overexpressed on tumour cells are selected and ligands specific to those receptors are conjugated on the surface of the MSNs. Specific drug delivery can be achieved by anchoring MSNs with targeting ligands. Folic acid (FA) is a well-known ligand which complements the folate receptors overexpressed on tumour cells. The FA conjugation on the surface of MSNs is brought about by an amide linkage between the carboxyl group of FA and amine group of aminopropyltriethoxysilane (APTES). Ma et al. delivered 5-aminolevulinic acid for photodynamic therapy by conjugating folic acid on the surface of HMSNs against B16F 10 skin cancer cells. The developed formulation was observed to show a high photocytotoxicity to the cancer cells. Similar reports were presented wherein the presence of amine functionalization on the surface aids in the binding of folic acid covalently on the surface of the receptors to ensure selective uptake of doxorubicin (DOX) in breast cancer cells. The apoptosis and cellular uptake studies revealed that FA-MSN-NH_2-DOX showed higher internalization into the cells as compared to MSN-NH_2-DOX . Hyaluronic acid (HA) is another widely explored ligand for targeting CD44 receptors overexpressed on cancer cells. Zhang et al. reported an HA functionalized DOX-MSN which behaved both as enzyme responsive and receptor-mediated delivery system. The increased uptake by colon cancer cells and low in vivo toxicity to the body was proved by in vivo tumour growth inhibition and biodistribution studies. In addition to receptor-mediated uptake, the release of DOX was triggered by hyaluronidase enzyme present in tumour microenvironment. Similar results were reported by Gary-Bobo et al. who fabricated HA-MSNs for photodynamic therapy against colon cancer. The experiments were conducted on HCT-116 colon cancer cell lines which showed the higher efficiency of HA-MSNs owing to their CD44 receptor targeted action as compared to that of plain MSNs.

Another novel approach for active targeting of hepatoma cells was reported wherein lactosaminated (Lac) MSNs were designed. These novel delivery systems showed a promising outcome as it had the ability to be endocytosed by asialoglycoprotein (ASGPR) receptors present on the surface of hepatocytes. The findings were supported by the results from cellular uptake studies in ASGPR-positive cells (HepG2 and SMMC7721) wherein Lac-MSNs showed improved cellular uptake when compared to that of plain MSNs. The results also highlighted an interesting point wherein the ligand lactose was recognized only when conjugated with MSNs and not in its free form . Analogues for targeting mannose-6-phosphate receptor overexpressed in cancer cells were grafted onto MSNs to enhance the uptake by tumour cells. These were found to show positive results for the treatment of prostate and colon cancers. Arginine-glycine-aspartic acid (RGD) was demonstrated to be selectively engulfed by $\alpha v\beta 3$ and $\alpha v\beta 5$ integrin receptors, which are overexpressed in diverse tumours. The drug-loaded surface engineered RGD-MSNs were thwarted by the normal cells and taken up by the liver cancer cells thus improving the treatment efficacy. In addition to molecular targeting, efforts were also made to club positron-emission tomography (PET) imaging and chemotherapy into a single carrier. Sunitinib as a model anticancer drug was loaded onto the MSNs and surface engineered with cyclo-(Arg-Gly-Asp-D-Tyr-Lys) peptide (cRGDyK) and polyethylene glycol. The efficacy of the receptor uptake was confirmed by flow cytometry studies, PET imaging and histopathological studies. The studies were carried out in U87MG human glioblastoma cells and athymic nude mice were used for in vivo experiments. The tumour uptake of the nanoconjugates was found to be lower in the case of plain HMSNs as compared to that of cRGDyK conjugated HMSNs. Numerous positive outcomes have been demonstrated using a plethora of ligands which are grafted onto the surface of MSNs rendering it target specific. Table lists a few of the ligands which have been explored for cancer drug delivery using MSNs.

Table: List of some functionalized MSNs explored for tumour targeting.

Drugs	Application	Targeting Ligand	Receptor
5-fluorouracil	Colorectal cancer	Hyaluronic acid	CD44
5-fluorouracil	Colorectal cancer	EGF	EGF
Curcumin	Cervical cancer	Chondroitin sulphate	CD44
Docetaxel	Breast cancer	Folic acid	Folate
Docetaxel	Hepatoma	Lactose	Asialoglycoprotein
Doxorubicin	Hepatic cancer	Transferrin	Transferrin
Doxorubicin	Colon cancer	Aptamer	(EpCAM)
Photosensitizer merocyanine	Breast cancer	Mannose	Mannose
Quercetin	Triple negative breast cancer	cRGD peptide	Integrin receptor αvβ3
Quercetin	Breast cancer	Folic acid	Folate
Sunitinib	Glioblastoma	$VEGF_{121}$	VEGF
Topotecan	Triple-negative breast cancer	cRGD peptide	Integrin receptor αvβ3

The surface functionalization of MSNs is not limited to just one ligand. Multiple ligands can be anchored onto the surface of MSNs for receptor targeting. For example, the surface of MSNs loaded with chlorambucil was functionalized with HA and RGD peptide for a synergistic effect. The in vitro cell line studies on human ovarian cancer (SKOV-3) cells showed a significant improvement in the uptake of the MSNs by CD44 and integrin receptors as compared to that of single ligand or plain MSNs.

In the recent times, a novel chemotherapeutic strategy utilizing copper impregnated MSNs (Cu-MSNs) have been explored for their potential reactive oxygen species (ROS) mediated apoptosis of cancer cells. Kankala et al. developed a novel Cu-MSN loaded with a catalase inhibitor, 3-amino-1,2,4-triazole (AT) for ROS mediated killing of cancer cells. On uptake by cells and delivery to the endosomal compartment (pH 5.0), the binding between copper and AT breaks off which releases AT to inhibit catalase activity. This in turn, leads to an increase in the ROS production induced by catalysis of copper on MSNs which aids in the cancer cell apoptosis. Positive results were obtained by them when evaluated in HT-29 cells. An effort to overcome the multidrug resistance observed in chemotherapy was made utilizing Cu-MSNs loaded with DOX as model drug. In this system, DOX was conjugated to copper metal through a pH sensitive coordination link susceptible to acidic tumour environment (pH 5.0–6.0). This entire system was further coated with liposomes comprising of cholesterol, d-α-tocopheryl polyethylene glycol 1000 succinate

(TPGS), MPEG-2000-DSPE (1,2-Distearoyl-phosphatidylethanolamine-methyl-polyethyleneglycol conjugate-2000), a P-gp inhibitor. Copper ions play a synergistic role in enhancing the intracellular ROS levels which results in killing of cancer cells efficiently. Positive results were obtained by the group when the delivery system was tested in DOX-resistant tumour (MES-SA-DX-5 derived from human uterine sarcoma) and HT-29 cell lines. The antitumor activity of Liposomes-Cu-MSN-DOX was found to be higher when compared to that of pure DOX. The antiproliferative activity of the optimized formulation was found to be profoundly higher in drug resistant cells than the sensitive cells.

MSNs for Anti-inflammatory Activities

The unique properties of MSNs have been used to accommodate various anti-inflammatory drugs and control their release rate. MCM-41 and SBA-15 are the two widely used silica drug carriers. Ibuprofen was loaded onto multimodal pore channels and the effect on its release was studied. On analysing the release data, it was observed that the release of the drug occurs in three stages viz, initial rapid release of the drug adsorbed on the surface of MSNs followed by prolonged release of the drug entrapped in the small pores present in the periphery of the particles having a different pore structural orientation compared to the inner mesopores. The final prolonged release stage occurs by the drug embedded deep in the long length pores of the particles. The bulk of the research work on MSNs for effective delivery of anti-inflammatory drugs has been carried out using ibuprofen as a model drug. Efforts were made to modulate the release rate of the drug by various organic modifications of the surface silanol groups. Surface functionalization was found to alter and also stabilize the drug release rates. SBA-15 has also been used to precisely deliver ibuprofen and other anti-inflammatory agents. One such work demonstrated a change in the release rate of ibuprofen when the surface of the carrier was modified with aminopropyl groups. Plain SBA-15 was not able to sustain the release of the drug due to weak interaction between the surface silanol group and the carboxyl group of ibuprofen. However, on amino functionalization, a relatively prolonged release of the drug was observed due to the strong interaction between the amino groups of silica surface and carboxyl groups of ibuprofen. Other similar studies were carried out to successfully deliver aspirin and indomethacin using MSNs with high drug loading and slow release profile.

Gated Drug Release/Controlled Drug Delivery

Although the length and the pore structure altered the drug release rate, efforts are constantly underway to achieve smart, zero-release of the drugs by capping the surface of the pores. Controlled and intelligent delivery of drugs to the target site through MSNs is possible due to gated release. The gates of the pores open only in response to certain stimuli like pH, temperature, enzyme, redox and so forth. The principle of gated drug release is very effective when toxic side effects of drugs to other organs are to be avoided.

pH-responsive Drug Release

pH is the widely explored stimuli to trigger the drug release as the body has a wide range of pH. MCM-48 particles loaded with prednisolone were coated with succinylated ε-polylysine (SPL) to ensure the pH-dependent release of the drug in the colon region. The in vitro release experiments showed a delay in the release of drug which indicated the successful approach of pH-responsive drug delivery. At acidic pH of the stomach, SPL prevents drug release due to its unionized form whereas, at colonic pH, SPL gets converted to its ionized form facilitating drug diffusion out of the MSNs. The developed nanoparticles could be an alternative for the treatment of diseases of the colon (inflammatory bowel disease and cancer). The promising outcome of pH-responsive MSNs was observed for tumour-targeted therapy and another disease where the pH of the affected area is slightly acidic than the normal tissues. This pH-responsive drug release can be realized by capping the pores of MSN using acid degradable polymers, polyelectrolytes, some pH-sensitive linkers and so forth. pH sensitive polysaccharide, chitosan was coated onto the MSNs to achieve controlled delivery of curcumin for the treatment of cancer. In vitro drug release studies proved the pH-sensitive nature of chitosan by sustained drug release. The release of curcumin improved when moving from pH 7.4 to pH 5.5. Cell uptake studies in U87MG glioblastoma cancer cell-line showed a decrease in the half maximal inhibitory concentration (IC50) values indicating an improved accumulation of curcumin in cancer cells when encapsulated within chitosan loaded MSNs. Similar studies using chitosan as the pH-triggered cap was carried out by Hu et al. for the release of doxorubicin on MCF-7 breast cancer cells. At acidic pH, the amino groups of chitosan become protonated resulting in swelling of the polymer chains. This opens up the pores of MSNs releasing the drugs. Modulation of doxorubicin via pH-responsive stimuli was also achieved using polymers such as poly(acrylic acid) and polydopamine. Both these reports suggested an enhanced uptake by cancer cells and a sustained release of the drugs. Reports with different polyelectrolytes as pH motifs have shown promising outcomes. Tannins as pH motifs were proved by the work of Hu and collaborators. Tannins, by the formation of boronate esters were found to modulate the release of the dye rhodamine at acidic pH.

Macromolecular compounds like cyclodextrin (CD) has also been explored for its pH-sensitive property. Tan et al. used a stalk of p-anisidine, loaded the cargo and finally capped the pores with β-CD. The pH sensitivity of the resulting MSNs was evaluated over a pH range of 7.4 to 5.5. A pH-dependent release was observed with a greater release at pH 5.5. When the pH fell below the pKa of the stalk, the interaction between p-anisidine and β-CD reduced thus resulting in the release of the drug. An optimum density of the stalk was the key to control the pH-triggered release. In addition, the effect of both α and β-CD capping on the release of drug was studied. It was observed that the percentage release with α-CD was comparatively lesser that β-CD. This could be due to the different formation constants between CD and p-anisidine stalks. Figure illustrates the release of drug from β-CD capped MSNs in response to stimuli.

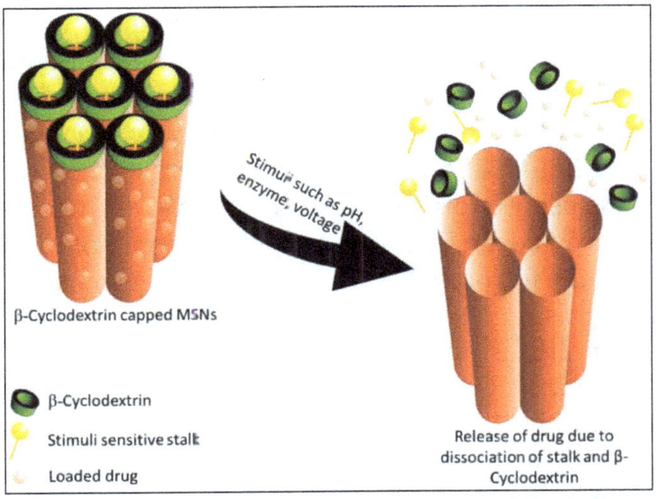

Schematic illustration of the release of drug from β-CD capped MSNs in response to stimuli.

Kuthati et al. devised a strategy to ensure the efficient delivery of antimicrobial agent by modifying the MSN framework via pH sensitive MSNs. Institute of Bioengineering and Nanotechnology-4 (IBN-4) nanoparticles, a type of MSNs was used as a carrier. This was immobilized with silver-indole-3-acetic acid hydrazide (IAAH-Ag) via a pH sensitive hydrazine bond. When exposed to the acidic environment of the site of bacterial infection (pH 5.0), the silver ions are preferentially released up to 12 h to ensure controlled release of the model drug by the cleavage of IAAH-Ag coordination bond. The antibacterial efficacy was investigated on two drug resistant strains of gram negative and gram positive bacteria namely Escherichia coli and Staphylococcus aureus respectively. In addition to this, the effect of IBN-4-IAAH-Ag NPs on inhibiting the formation of biofilms by four strains of bacteria viz, Escherichia coli, Bacillus subtilis, Staphylococcus aureus and Staphylococcus epidermidis revealed promising results. The in vitro results were substantiated by in vivo experiments on Escherichia coli infected C57BL6 mice. A remarkable reduction in Escherichia coli was observed in the formulation treated group of mice. Hence the developed system could be a plausible alternative to the antibiotics used currently.

Redox Responsive Drug Release

Redox triggering is another widely used strategy to control the release of cargo from carriers. This technique is being used to deliver anticancer drugs making use of endogenously present reducing agents. Redox cleavable disulfide bonds are often utilized for this kind of drug release. Wang et al. synthesized a disulfide-linked polyethylene glycol (PEG) tethered to MSN for redox responsive drug release. The efficiency of the synthesized MSNs was evaluated by in vitro release studies using Rhodamine B (RhB) as the model drug. Glutathione (GSH) equivalent to intracellular concentration was added to the release media. It was observed that in the absence of GSH, the release of RhB was negligible signifying the efficiency of the cap in blocking the drug release. In addition, the PEG surface modification conferred on the nanoparticles a significant biocompatibility. Dual responsive stalks with β-CD caps were synthesized to control the release of multiple drugs. HMSNs with acetal and ferrocene carboxylic acid units were obtained by selective functionalization. The redox responsiveness of the carrier was confirmed by in vitro experiments using hydrogen peroxide as stimuli. On increasing the concentration of hydrogen peroxide, an increase in the release rate of the drug was observed due to the oxidation of the ferrocenyl moiety. The mechanism behind the release is the strong electrostatic repulsion between β-CD and the oxidized hydrophilic ferrocenyl moiety which results in the dissociation of β-CD caps thus releasing the drug from the pores. The same carrier was also made pH-responsive by introducing an acetal linker to develop a synergistic effect.

Temperature Responsive Drug Release

Thermoresponsive MSNs have also been widely been studied as a possible means of controlling drug release. In this context, PEO-b-poly (N-isopropylacrylamide) based copolymeric micelles as structure directing agents for the synthesis of functionalized MSNs was developed by Bathfield et al. Ibuprofen as a model drug was loaded into the mesopores using one-pot strategy wherein the drug was incorporated directly into the hybrid material. Ultimately, the structure directing agent in this formulation was the drug-loaded polymer micelles. The drug release profile at 20 and 45 °C revealed a temperature sensitive pattern with a higher drug release at 45 °C than that at 20 °C.

Chemical and Enzyme Responsive Drug Release

Several chemicals and enzymes present inherently in the body or produced during diseased conditions have also been explored for the possibility of triggering drug release from the MSNs. Glucose as a chemical has been studied by scientists as a possible trigger to release drugs and has been a boon for diabetes management. In one of the studies, MSNs functionalized with a signal reporter, alizarin complexone (ALC) was developed. Gluconated insulin was3 then introduced within the pores by benzene-1,4-diboronic acid (BA) mediated esterification reaction. This behaved both as a hypoglycaemic agent as well as pore blocker. In addition, rosiglitazone maleate was also introduced into the pores to form multifunctional MSN. In the presence of glucose, competitive binding between ALC and BA occurs which leads to opening of the pores and release of the drug. They fabricated glucosamine-poly(acrylic acid) conjugated MSNs. The pores were capped by the crosslinking of the poly(acrylic acid) chains by the formation of boronate esters. The drug release was governed by pH or the presence of glucose. At mild acidic conditions of pH 6.0 and the presence of 10 mM glucose, the drug release was found to be higher by around 65%. These results suggested that at pH 6.0, combined stimuli showed a sufficiently enhanced release of the drug. Another novel chemical sensitive MSN was with that using thrombin. The MSNs were loaded with an anticoagulant drug (acenocoumarol) and the pores were capped with peptide LVPRGSGGLVPRGSGGLVPRGSK-pentanoic acid (P) which is a substrate for proteolytic α-human thrombin. The results demonstrated that the release of drug was highly specific to the presence of thrombin. Thrombin results in the hydrolysis of the capping peptide releasing the drug.

Tailoring the drug release in response to the presence of certain enzymes is another new avenue for modifying the MSN as a carrier. NAD(P)H: Quinone oxidoreductase 1 (NQO1) enzyme as stimuli for the release of doxorubicin was demonstrated by Gayam et al. They synthesized MSNs surface functionalized successively with alkyne followed by drug loading. To avoid premature drug release, the mesopores were blocked with rotaxane followed by tethering it to benzoquinone. In the presence of NQO1 enzyme which is upregulated in several tumours, the benzoquinone gets reduced leading to the opening of the pores and the release of doxorubicin. In vivo experiments in nude mice bearing tumour (lung) showed promising results with a significant reduction in tumour volume in the mice treated with MSN-NQO1 compared to those treated with saline and free doxorubicin. Matrix metalloproteinase (MMP-2) triggered drug release for the treatment of liver cancer is another possible line of therapy. Many tumours have shown to overexpress MMP-2, the advantage of which could be explored to target drugs to tumour sites. A two-component that is, a cell penetrating peptide polyarginine and PVGLIG which is a substrate for MMP-2 cleaving based polypeptide was linked with phenylboronic acid-human serum albumin (PBA-HSA) onto the MSNs to render it target specific. Doxorubicin (DOX) was used as the model drug. Human serum albumin was used to cap the pores and phenylboronic acid behaved as the targeting moiety specific to sialic acid overexpressed in liver tumours. The enzyme responsive behaviour of the carrier was studied in vitro where around 73% of the drug released in the presence of MMP-2 as compared to only 15% of release in the absence of MMP-2.

In vivo studies in HepG2 cells injected nude mice were performed which revealed encouraging results with DOX-loaded MSN-HSA-PBA showing significant tumour growth inhibition compared to non-functionalized MSN loaded with DOX . An experiment along similar line was performed by Radhakrishnan et al. who explored the use of protamine, a peptide drug as a capping agent to prevent the premature re-lease of diclofenac from MSNs. The protamine cap was found to get hydrolysed in the presence of trypsin enzyme which cleaves the l-arginine residues in protamine thereby releasing the drug. This concept was evaluated by studies on COLO 205 cells. The re-sults showed that about 87% of the drug released within 120 min in COLO 205 cells as compared to 13% in healthy cells. This shows the selective drug release in colon cancer cells as compared to normal cells. Radhakrishnan et al. also demonstrated the use of chondroitin sulphate (CHD) as a gate to the pores of MSNs which would trigger the release of drug in the presence of hyaluronidase enzyme. In addition, CHD also acts as a ligand which gets specifically uptaken by cells overexpressing CD44 receptors. CHD behaves as a cap which prevents the outward diffusion of the drug in the absence of hyaluronidase enzyme.

In one of the studies, controlled release of drug was achieved by surface functionalization of MSNs with ferrocenyl moiety β-cyclodextrin complex (Fc-β-CD) to prevent premature release of the drug. The experiments suggested that on stimulation by heme protein (horse-radish peroxidase and hydrogen peroxide) and the production of hydrogen peroxide via the oxidation of glucose by glucose oxidase or +1.5 V stimuli, the ferrocenyl moiety gets dissociated resulting in the opening of the nanovalves and thus releasing the drug. In vitro studies conducted to assess the release property of the carrier revealed that on stimulation with heme protein, a greater amount of drug was released as compared to that with glucose stimulation and without stimuli.

External Stimuli for Drug Release

Various other stimuli have also been investigated in modulating the drug delivery from MSNs such as light, magnetic, ultrasound, electroresponsive systems. Figure depicts the release of drug from gated MSNs in response to stimuli.

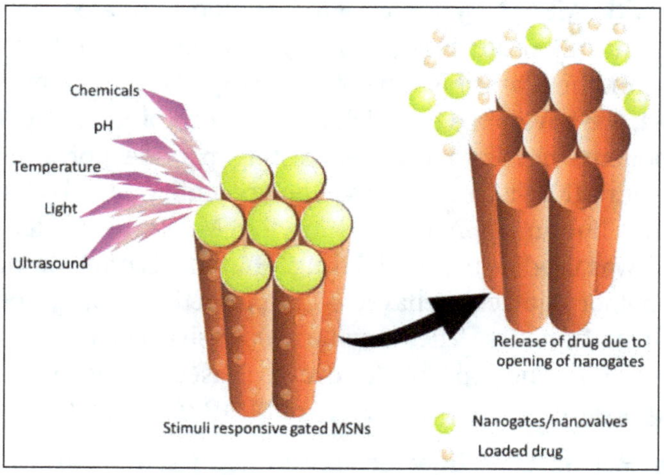

Schematic illustration of the release of drug from gated MSNs in response to stimuli.

Use of light as an activating mechanism of drug delivery has garnered attention due to its advantages of spatial and temporal control of release of drugs. Febvay et al. combined the advantages of light triggering with a high loading of MSNs in their study to enable the delivery of a model molecule, a fluorescent dye–Alexa546–which is impermeable to the cytosolic compartment. The MSNs were prepared using secondary surfactant pluronic F127 and further tagged with biotin or streptavidin. Following endocytosis, the LN-229 cells (human glioma, ATCC) were exposed to green excitation light for 3 to 120 s which resulted in the dye being released into the cytosol. This could be attributed to the cell membrane damage induced by the light irradiation. Liu et al. explored the potential of deeper tissue penetration potential of near infrared (NIR) radiations to control the release of the drugs. They developed MSNs with gold nanorods forming the inner core and phase changing molecule, 1-tetradecanol as gatekeepers.

DOX was chosen as the model drug molecule. These nanoparticles were further functionalized with folate moieties to target KB cells. The system was able to release DOX due to heating induced by near IR radiation generated by a gold nanorod core. The successful release of the drug on IR radiation was ascertained by first subjecting it to external heating which did not show any release at 37 °C. On NIR irradiation of 802 nm for 10 min, the temperature increased to 45 °C which resulted in the release of DOX which may be attributed to the change in the fluid state of tetradecanol molecules above T_m.

In vitro studies on KB cells showed preferential uptake by the cell owing to the surface functionalization. The combined effect of phototherapy (NIR irradiation), chemotherapy (DOX) and targeted therapy resulted in remarkable killing of cancer cells. Hence the developed nanoparticulate system could be efficiently used for multimodal therapy. Li et al. worked on similar lines where they explored the potential of MSNs as multimodal carriers using MRI monitored magnetic targeting and NIR mediated phototherapy. In their work, they developed iron oxide (Fe_3O_4) nanoparticles coated with trisoctahedral gold (Au) shell which were loaded with DOX. These were further coated with silica shell capped with oligonucleotides. Au shell here acted as NIR responsive material whereas Fe_3O_4 facilitated magnetically triggered release of drug. The controlled release property of the carrier was confirmed via in vitro studies on HeLa cell lines which showed DOX release upon NIR irradiation at 808 nm and magnetic attraction for a brief period of 2 h. These results were further supported by in vivo studies on nude mice. HeLa cells were transplanted to the nude mice for the development of tumour after which the nanoparticles were injected. On exposure to magnetic attraction for 30 min and laser exposure at 3 W/cm^2 for 30 min, a complete disappearance of tumour was observed after 14 days. These findings suggest the plausible use of these carriers to enable efficient combination therapy options for tumours.

Another novel strategy to induce visible light irradiated delivery of drugs using MSNs as drug carriers was put forth by Kuthati et al. They developed a silver nanoparticle (SNPs) decorated copper impregnated MSNs (Cu-MSNs) to aid in the photodynamic inactivation of antibiotic resistant Escherichia coli. Curcumin (Cur), a phototherapeutic agent was loaded into the MSNs. They explored the plasmonic resonance coupling between curcumin and silver nanoparticles to enhance the transfer of silver nanoparticle to photosensitizer to effectively kill gram negative Escherichia coli. They hypothesized that curcumin would produce large amounts of ROS under light irradiation which will enhance the release of silver ions. SNP+Cur behaves as a positively charged nanocomposite which improves the binding ability of Cur to the bacterial membrane. The antibacterial efficacy against Escherichia coli was studied by constant illumination using LED array at 470 nm at a fluence of 72 J/cm^2. The study yielded promising results revealing that Cur-Cu-MSN-SNP efficiently killed bacteria in light conditions with very mild toxicity in dark conditions. The successful eradication of bacterial cells may be attributed to three factors, namely, improved solubility and local concentration of Cur loaded into MSNs, improved binding of positively charged nanoparticle to bacterial membrane, enhanced ROS production due to Cu-MSN-SNPs. Similar strategy can be used for the development of wide variety of photobactericidal systems.

The release of drugs from MSNs can also be triggered via magnetic attraction. Baeza et al. reported the synthesis of a hybrid polymer which responded to both thermal and magnetic stimuli. They incorporated superparamagnetic iron-oxide nanocrystals into the mesopores which was capable of providing a sufficient heating capacity for hypothermia cancer therapy. To enable thermoresponsive release of the drug molecule, poly(N-isopropylacrylamide) was tethered to the surface of MSNs. Fluorescein and soybean trypsin inhibitor type II was loaded into the pores of the MSNs to model the release property. In response to an alternating magnetic field of 24 kA/m and 100 kHz inside a thermostatic chamber maintained at 20 °C, a higher release of fluorescein was observed which may be due to the heat energy and enlargement of the pores leading to greater release of the drugs.

Among the non-invasive routes to enhance the spatio-temporal delivery of drugs, ultrasound is slowly gaining popularity owing to its advantages of lower cost, absence of ionizing radiations and ease of tissue penetration regulation by tuning the frequency of the cycles and exposure time. They are capable of inducing thermal/mechanical effects which can trigger the release of the drugs. Several efforts are underway to utilize this technology for the therapy of diseases especially cancer. Paris et al. formulated an ultrasound responsive MSNs capped with the copolymer, poly(2-(2-methoxyethoxy) ethyl methacrylate-co-2-tetrahydropyranyl methacrylate), with two functional ends and a monomer ratio 90:10 (MEO_2MA:THPMA). They utilized fluorescein as the model molecule to monitor the developed system. Ultrasound irradiation resulted in the cleavage of the hydrophobic tetrahydropyranyl moiety of the cap leading to a change in its conformation allowing the drug to be released. They studied this concept further by surface functionalization of the MSNs with biotin and RGD peptide to enable selective uptake by tumour cells (HeLa cell lines). DOX was loaded as the cargo within the mesopores. They successfully proved the efficiency of the developed nanoparticulate system in enhancing the efficiency of the DOX in killing cancer cells Studies on similar line was reported by Kim et al. using ibuprofen as a model drug which were loaded into MSNs and covered with poly(dimethylsiloxane) as the implantable body. On exposure to ultrasound of 28 kHz with a power of 1.5 W/cm^2, an increase in the diffusion of ibuprofen was observed. In addition, negligible damage was observed to the polymer material on subjecting it to ultrasound frequency. These results suggest the possible use of this stimulus in enabling controlled release of drug to the diseased sites.

Another interesting approach to non-invasive drug therapy could be the use of a mild electric field that aids in controlling the release of the cargo by activating certain mechanisms, viz, electrochemical reduction-oxidation and movement of a charged molecule. Xiao et al. developed a novel controlled release MSNs sensitive to enzyme or voltage. These were functionalized with ferrocene and further loaded with rhodamine and capped with β-cyclodextrin (β-CD). Voltage ranging from 0.5 V to 1.5 V was applied to trigger the release of the cargo from the pores. Both the enzyme and voltage triggered release was based on the presence of ferrocenyl (Fc) and β-CD valve. The release is based on the conversion of Fc to Fc^+ which dissociates from the surface of the pores under a standard potential of +0.32 V. They observed that at a higher voltage, the release rate was higher and vice versa. Wang et al. also utilized the potential of ferrocene functionalized β-CD as nanovalves to modulate the drug release from MSNs. They loaded two drugs that is, gemcitabine and doxorubicin and evaluated the potential of ferrocene in controlling the drug release. On applying a voltage of +1.5 V, a controlled release of 23% of gemcitabine was observed over 15 min. Thereafter −1.5 V was applied, which ceased the release of the drug. Hence, this approach can be effectively used for controlling the drug release from MSNs and thus avoiding side effects.

MSNs for Improvement of Solubility of Drugs

MSNs owing to their modifiable surface chemistry can act as carriers for poorly soluble drugs and tackle their solubility issues. Bukara et al. proved this potential of MSNs by loading the poorly soluble drug fenofibrate and assessing them in healthy human volunteers. The volunteers were monitored for a period of 96 h post dosing and their plasma samples were collected and assessed for the pharmacologically active metabolite fenofibric acid. A significant increase in C_{max} with a point estimate of 177% and a reduction in t_{max} were observed for fenofibrate formulation following single oral administration. No serious adverse events were reported and none of the volunteers discontinued the study. This demonstrates that the MSNs could also be used as a possible alternative to other carriers to improve the solubility and bioavailability of drugs. Enhanced oral bioavailability of telmisartan (TEL) was achieved by loading it into MSNs. Based on the results obtained by the study on beagle dogs, Zhang et al. set forth a basis to use MSNs as a drug carrier for poorly soluble drugs. In vitro cellular uptake studies showed that TEL-MSN showed an enhanced uptake in Caco-2 cells resulting in accumulation in the cell membrane as compared to TEL-mesoporous silica microparticles. The uptake mechanism of these MSNs occurred in three major steps: first, binding of MSNs to intestinal cell membrane followed by nonspecific cellular uptake and merging with endosomes and finally, release from endosomes and enters the cytoplasm. In vivo absorption studies of TEL-MSN in beagle dogs revealed a 1.29 times increase in $AUC_{0\rightarrow 72h}$ as compared to that of marketed tablet and MSN. Thomas et al. synthesized MSNs loaded with BSC-II antiepileptic drugs, carbamazepine (CBZ), oxcarbazepine (OXC) and rufinamide (RFN). The dissolution profile of the drugs in phosphate buffer, showed faster release for CBZ and OXC whereas RFN showed a slower release of drugs after an initial burst release. The profile resembled a first order release mechanism related to the drug diffusion process. This could be widely exploited for the improvement in drug absorption and bioavailability of poorly soluble drugs by further proving this concept via in vivo studies.

MSNs in Biomedical Imaging and Theranostic Purpose

The versatile features of MSNs such as ability to incorporate wide variety and large number of compounds within them, their stability and controllable size, makes it an ideal platform for biomedical imaging and theranostic applications. Many of the fluorophores face certain drawbacks such as poor solubility and stability especially those for near infrared (NIR) imaging. These compounds can be incorporated within the MSNs to improve their photophysical and photochemical properties. Various literatures have suggested the successful utilization of this carrier for imaging and theranostic purposes.

MSNs can be widely used for optical imaging, magnetic resonance imaging (MRI), positron emission tomography (PET). Optical imaging is a technique wherein the specific probes are excited by incident light usually in the visible or near infrared regions, thus emitting light at a lower energy. MRI is a powerful in vivo imaging technique which gives a three dimensional anatomical picture of the region of interest with a high resolution.

Most often, the dyes are incorporated within the mesopores which gives sufficient stability and protection from the external environment. Sreejith et al. developed a novel hybrid material constituting squaraine loaded MSNs which were further coated with thin sheets of graphene oxide. Squaraine dyes possess significant photophysical properties in the NIR region. In order to ascertain their bioimaging ability, in vitro studies were carried out on HeLa cell lines which showed positive results. The developed hybrid formulation was successful in protecting the dye as well as preventing its leakage from the system showing a potential platform for bioimaging. Moreover, these MSNs can also be tagged with surface active moieties which can be preferentially directed to abnormal tissues for diagnostic and therapeutic purposes. Nakamura et al. reported the synthesis of multimodal MSNs possessing features of imaging as well as drug delivery. The carrier was loaded with ^{19}F, a MRI contrast reagent. These nanoparticles were further labeled with fluorescent dyes and tethered with folic acid to ensure adequate uptake by tumour cells. The developed nanoparticles demonstrated positive results in vitro exhibiting sufficient cellular uptake by folate expressing cancer cells which was observed via ^{19}F MRI and fluorescence microscopy. Jun et al. reported the use of silica nanoparticles embedded with quantum dots (QDs) and constituting a core-shell of CdSe@ZnS for bioimaging purposes (Si@QDs@Si NPs). They observed that the system showed superior fluorescence as compared to single quantum dots when studied in HeLa cells. The same was confirmed via in vivo testing wherein the mice were injected with Si@QDs@Si NPs mixed with HeLa cells. The nanoparticles exhibited enhanced fluorescence and hence can be effectively used for bioimaging which requires minute cell tracking with high sensitivity. Helle et al. explored the potential of cyanine 7-doped silica nanoparticles for lymph node mapping using NIR imaging. They were able to map the lymph nodes for the diagnosis of possible metastatic and draining nodes. In addition, the developed platform were found to be excreted via hepatobiliary route and was found to be safe when tested in mice up to a period of three months showing an efficient hepatobiliary excretion.

MSNs for Bone Tissue Engineering and Repair

MSNs have found a special place in the field of tissue engineering. Most of the research in this area revolves around bone tissue differentiation and osteogenesis. The surface silanol groups present on MSNs react with the body fluids to generate carbonated apatite which can further lead to bone generation. In addition to this, the MSNs can be loaded with osteogenic agents to augment the bone tissue engineering process. For instance, bone morphogenetic protein-2 (BMP2) derived peptide functionalized dexamethasone loaded MSNs were formulated to evaluate its efficacy for the osteogenic differentiation. The evaluation was carried out using cell line studies to study the endocytosis and uptake of the functionalized MSNs. The ectopic bone formation was studied in vivo, the results of which indicated that the BMP2-pep functionalized MSNs held great promise in bone repair. The addition of dexamethasone synergized the bone differentiation effect. Similar results were reported by Luo et al. for bone forming peptide incorporated MSNs. Incorporation of bioactive glasses into mesoporous silica is another interesting application of MSN in bone repair. Increased bone tissue regeneration was observed with these materials containing SiO_2-CaO-P_2O_5 as composition. Readers are directed to references for a detailed review of these materials.

Biodistribution and Biocompatibility of MSNs

The safety and toxicity of nanoparticles are a cause of major concern owing to their high surface-to-volume ratio compared to its counterparts. The biocompatibility of any carrier is a prerequisite property for any pharmaceutical product to ascertain that these products do not accumulate in the body over a period of time causing untoward effects.

Many of the formulations containing conventional nanocarriers have been approved by US FDA. Biomaterials such as lipids and polymers constitute these conventional nanocarriers. Due to its inherent biodegradability and biocompatibility, these nanocarriers have been constantly exploited for further research to enhance its biomedical applicability. Liposomes are one such carrier which comprises of phospholipid bilayer within which both hydrophobic and hydrophilic drug can be encapsulated. Liposomes have proven to be capable of being used for site specific drug targeting in a variety of diseases. These carriers have been found to be safe which may be attributed to the biocompatible nature of phospholipids used. Another nanocarrier that shares a similar importance to that of liposomes is the polymeric nanocarrier. In this regard, poly(-lactic-co-glycolic acid) (PLGA) is one of the renowned polymer-based carriers. This is a part of an FDA approved device. Although these encouraging results have been obtained regarding their safety for human use, there are certain drawbacks such as stability related issues, lack of control over drug release and difficulty in overcoming certain biological barriers associated with these carriers. Nevertheless, these are still the most widely explored carriers due to their non-toxic property.

Inorganic nanocarriers with robust characteristics, MSNs, although has shown positive in vitro results, studies are still being carried out extensively due to the age-old toxicity-related issues of silicon dioxide, especially silicosis. Efforts are underway in identifying the major routes of toxicity of silica in both its crystalline as well as amorphous forms.

Effect of Surface Chemistry, Shape and Size of MSNs

As per the reports, the major pathway of toxicity associated with silica is due to its surface chemistry (silanol groups) which can interact with the membrane components leading to the lysis of the cells and leaking of the cellular components. Mesoporous silica exhibited lower hemolytic effect compared to non-porous silica. This could be attributed to the lower density of silanol groups on the surface of mesoporous structures. Surface properties of MSNs also have a great impact on the biodistribution and biocompatibility of MSNs. Altering the surface features by functionalization with polyethylene glycol (PEG) helps the MSNs to escape from being captured by liver, spleen and lung tissues. This could be attributed to the longer circulation time of PEG-MSNs. Yu et al. studied the impact of pore size, shape and surface features of silica nanoparticles on the cellular toxicity. The cellular toxicity was evaluated on macrophages (RAW 264.7) and cancer epithelial (A549) cells. Post 72 h exposure, they observed that A549 cells were resistant to the nanoparticles even at the concentration of 500 μg/mL. However, at a concentration of 1000 μg/mL, observable toxicity was seen. The IC50 value for the nanoparticles when tested on macrophages was found to be between 50 and 100 μg/mL. The cellular level of association was determined using inductively coupled plasma mass spectrometry (ICP-MS). Interestingly; it was observed that amino modified MSNs showed a higher level of cellular association which is contradictory to literature which report that increase in surface silanol groups are responsible for higher cellular association. The plausible explanation for this higher interaction between amino MSNs and cells could be that a particular surface threshold exists beyond which cell interaction is facilitated. The observations suggest that toxicity depends on the type of cells, concentration of nanoparticles treated and pore size and surface charge of nanoparticles.

An interesting experiment to identify the effect of the spatial arrangement of MSN surface amine groups on its interaction pattern with cells was performed by Townson et al. MSNs with the same size, porosity and charge were modified with suitable reagents (trimethoxysilylpropyl modified polyethyleneimine, 2-[methoxy(polyethyleneoxy)propyl]trimethoxysilane, N-trimethoxysilylpropyl-N,N,N-trimethyl ammonium chloride) resulting in PEG-PEI and PEG-NMe_3^+ MSNs to ensure exposed polyamines and distributed, obstructed amine groups respectively. Both in vitro and in vivo experiments were performed to determine the toxicity effects. The synthesized nanoparticles were subjected to cytotoxicity studies on a wide range of cell lines such as A549 (human lung carcinoma), A431 (human epithelial cancer), Hep3B (human hepatocellular carcinoma) and human hepatocytes. PEG-PEI MSNs were found to bind to all the cells whereas PEG-NMe_3^+ MSNs showed limited binding. To confirm these results and to ascertain if the same effects will be observed in a biological system as well, 50 μg was injected into the veins of ex vivo chick embryos which helped in real-time imaging of particles. The results showed a similar trend as that observed in in vitro studies. PEG-PEI MSNs were found to bind endothelial cells and stationary and circulating white blood cells (WBCs) whereas PEG-NMe_3^+ MSNs remained in circulation for >6 h. In order to verify the importance of amine groups in binding, the MSNs were subjected to acetylation thereby shielding the amine groups. This reduces the binding affinity of the MSNs. it was concluded that exposure to charged particles and its effect on the formation of protein corona in vivo should also be considered when designing MSNs for biomedical applications.

The biodegradation and toxicity of MSNs also depend on the shape of the MSNs. The effect of shape on in vivo toxicity of MSNs after oral administration was studied by Li et al. for MSNs with different aspect ratios of 1, 1.75 and 5. These MSNs were administered at a dose of 40 mg/kg. Two hours post administration, a significant number of MSNs were observed in the liver and spleen. The number of spherical nanoparticles showed a marked increase in the liver compared to its rod-shaped counterparts. It was observed that MSNs showed a rapid excretion from the body via faeces while some of the unchanged MSNs or their degradation products could be absorbed and later excreted via urine. The in vitro results of degradation showed that spherical nanoparticles showed rapid degradation while long rods have slow degradation rate, especially in intestinal fluid. These results suggest that degradation of MSNs depends on the shape and biological environment. No abnormalities were observed in liver, spleen, lung and heart. However, spherical nanoparticles induced renal tubular necrosis and haemorrhage which may be due to the degradation products. Contrasting results were obtained for MSNs administered via an intravenous route where no major abnormality in kidney was detected. However, there is a lack of clear picture of the degradation pathway and pharmacokinetics of MSNs. With this regard, efforts were made by Zhao et al. to study the pharmacokinetics and biodistribution of different shapes of MSNs namely, long rod, short rod and spherical particles following oral administration. The retention of the nanoparticles in the gastrointestinal tract was determined by ex vivo optical imaging method. In vitro cytotoxicity study revealed that all the three nanoparticles were nearly non-toxic in nature which can be attributed to the conversion of nanoparticles to non-toxic silicate ions. To predict the biodistribution of the nanoparticles, Si content in different organs was determined by inductively coupled plasma optical emission spectrometry (ICP-OES). On examining the Si content in different organs, Zhao et al. also arrived at the same conclusion as that by Li, that majority of the MSNs accumulate in the liver. The synthesized MSNs did not show any visible histopathological changes when compared with that of the control indicating that the particles did not produce any gastrointestinal toxicity or inflammation.

Size of the nanoparticles also has a profound influence on the biodistribution and excretion of MSNs. MSNs with a varying particle size from 80 to 360 nm was prepared and their biodistribution was assessed in ICR mice. An increase in particle size led to an increase in its accumulation in the liver and spleen following intravenous (i.v.) administration. However, no pathological abnormalities were observed at the end of 1 month. Smaller sized MSNs undergo slower degradation as it can escape the degradation by liver and spleen. Different types of MSNs (MCM-41, SBA-15) were injected subcutaneously (s.c.) and intraperitoneally (i.p.) to mice. Intraperitoneal injection resulted in the death of animals which may be attributed to the rapid systemic distribution following i.p. injection as compared to s.c. Acute and sub-acute toxicity profiling of fluorescent mesoporous silica nanoparticles (FMSNs) were performed in female nude mice. 1 mg/mouse/day was administered via the intravenous route. No observable toxicity was seen in the animals. Long-term toxicity study following intraperitoneal injection of FMSNs at a dose of 1 mg/mouse/day twice per week for 2 months was conducted to assess the long-term effects of MSNs. The histopathological examination of body tissues, haematological parameters displayed no apparent changes compared to control. In addition, the FMSNs also showed an enhanced tumour uptake property resulting in a reduction in tumour volume. Single dose toxicity studies by Tang et al. revealed that the nanoparticles exhibited a size-dependent toxicity. Zhang et al. synthesized DOX-loaded MSNs functionalized with folic acid of varying sizes of 48, 72 and 100 nm and investigated the effect of particle size on its in vivo distribution in MDA-MB-231 tumour-bearing Balb/c mice. The animals were sacrificed at the end of 24, 48 and 72 h post injection and their organs were harvested. The amount of Si content in each of the organs was determined by inductively coupled plasma mass spectrometry (ICP-MS). It was observed that MSNs with size of 48 nm showed the highest accumulation in the tumour tissues. The results suggest that particle size and surface modification alters the biodistribution of MSNs.

The data generated from various literatures suggest that careful control of particle size and shape is the determinant factor in ascertaining the biodistribution and toxicity of MSNs. In addition, the safety and toxicity of MSNs also depend on the dosage of the MSNs administered at which no observable biological effects are detected.

Recently, Shen et al. reported that the novel 3D-dendritic MSNs synthesized by them showed a rapid biodegradation in simulated body fluid within 24 h as compared to two weeks for that of plain MSNs reported earlier. In yet another effort to prepare biodegradable MSNs, He et al. synthesized a novel pH responsive mesoporous silica–calcium phosphate (MSN-CAP) hybrid nanoparticles. The MSNs were doped with calcium phosphate during synthesis process which yielded pH responsive MSNs. The in vitro degradation behaviour of MSN-CAP was observed in simulated fluids. The results showed that the complete degradation of the nanoparticles took place in 24 h. Both these novel MSNs could be an alternative prospect for clinical use. However, their in vivo degradation behaviour has to be ascertained.

In Vivo Safety and Toxicity of MSNs

Determining the safety and biocompatibility of MSNs is crucial owing to its variable characteristics. The toxicity of the carriers depends on its various characteristics and the conclusions derived from the studies were found to vary. Nevertheless, most of the reports showed that the MSNs get preferentially accumulated in the liver and spleen following administration.

Liu and collaborators made an attempt to study the single and repeated dose toxicity of HMSNs following intravenous administration in mice. LD_{50} of HMSNs was found to be higher than 1000 mg/kg. In single dose toxicity studies, mice were injected with HMSNs at a low dose and high dose. At the higher dose of 1280 mg/kg, mice did not survive. In contrast, the groups treated with low dose HMSNs did not show any behavioural changes nor any haematology or pathological changes. To carry out the detailed repeated dose toxicity studies, intravenous administration of HMSNs were given to mice continuously for 14 days and observed for a month. During the 1 month observation period, no mortality was observed. Moreover, no remarkable changes in pathology or blood parameters were observed. In order to assess the fate of the nanoparticles, HMSNs were injected intravenously at a dose of 80 mg/kg. Following administration, the majority of the nanoparticles were found to localize in the liver and spleen. Analysis of the silicon content using ICP-OES revealed that highest amount of silica was present in the spleen and liver which gradually reduced over a period of 4 weeks.

In order to assess the fate of MSNs after different administration routes, Fu and collaborators tested MSNs with a particle size of 110 nm in ICR mice. Following administration via hypodermic, intramuscular and intravenous injection as well as oral administration, the in vivo distribution of fluorescent-tagged MSNs was tracked. It was observed that of all the exposure routes, the oral route was found to be well tolerated even when the dose was increased to 5000 mg/kg and intravenous route seemed to have the least threshold. MSNs administered via intravenous route were found to preferentially accumulate in the liver and spleen at the end of 24 h and 7 days whereas those administered by other routes did not show any fluorescence in these organs. It was observed that a portion of the MSNs administered via intramuscular and hypodermic route could cross different biological barriers with a slow absorption rate. The major routes of excretion of MSNs were found to be via urine and faeces with the highest values after oral administration as compared to other routes. No histopathological changes were observed in liver, spleen, kidney and lung at the end of 24 h and 7 days by different exposure routes. Nonetheless, a low degree of inflammation was seen in the mice which were treated with MSNs via the hypodermic and intramuscular route. The results suggested that MSNs were found to be safe and well tolerated when administered by oral and intravenous routes. For an extensive review on the biocompatibility of MSNs and silica NPs, the readers are directed to references.

MSNs vs. Silica Nanoparticles

Different forms of silica viz, fumed silica, porous silica and non-porous silica can be used as drug carriers. These carriers have shown encouraging results in preclinical studies. However, to translate these materials to the bedside, a clear understanding of the fate and the inherent toxicity of these carriers in vivo is essential. Silica nanoparticles used for biomedical applications are usually amorphous in nature belonging to either porous or non-porous category. The rapid clearance of amorphous silica from the lung compared to the crystalline forms is responsible for its lower toxicity potential. MSNs are found to dissolve rapidly when it is sufficiently below the saturation levels. As per the reports of Martin, silica dissolves in the body fluids which subsequently gets absorbed or excreted as silicic acid in the urine. The silica nanoparticles undergo degradation to silicic acid which is non-toxic via three different processes viz, hydration, hydrolysis and ion-exchange. This process of degradation was found to depend on the degradation medium and the concentration of nanoparticles. Various strategies have been explored which can manipulate the degradation kinetics of silica nanoparticles. Some of the approaches are noncovalent doping of organic moieties to accelerate the hydrolytic degradation, covalent binding of organically bridged silsesquioxanes- based NPs, and incorporation of cleavable organically bridged silsesquioxanes into silica NPs to enhance the degradation by the biological trigger. The degradation of MSNs is much more complicated than other silica NPs owing to its varying matrix. This is attributed to the difference in the rate and degree of condensation of silica matrices between the various sol-gel procedures of MSN synthesis. As per the review by Croissant et al., partially condensed MSNs degrade in a few days, well condensed MSNs tend to degrade in weeks and calcined MSNs takes months for its degradation.

In order to assess the impact of porous structures on the in vivo immunotoxicity, Lee et al. performed repeated-dose toxicity studies on BALB/c mice. MSNs and colloidal silica NPs were injected intraperitoneally into mice for 4 weeks. At the end of the study period, the animals were sacrificed and organs were harvested to study the effects of the particles. The animals treated with MSNs showed an increase in the relative weight of liver and spleen and an increased response to lymphocyte mitogens, concanavalin A (Con A) or lipopolysaccharide (LPS). In addition, a decrease in the $CD4^+/CD8^-$ and $CD4^-/CD8^+$ phenotypes and an increase in the levels of $CD4^+/CD8^-$ and $CD4^+/CD8^+$ were recorded. Elevated IgG/IgM levels were also observed in the MSN treated animals. The results indicate that MSNs showed a greater extent of damage than colloidal silica NPs However, a careful study of the toxicity is needed to establish the safety of MSNs.

Biocompatibility of MSNs in Humans

The biocompatibility of silica nanoparticles has long been a topic of controversy as studies conducted by researchers have yielded variable results. Nevertheless, the Food and Drug Administration (FDA) approval of hybrid silica nanoparticles for bioimaging marks an event of utmost importance. These particles were found to be ~7 nm in size within which fluorescent dye, Cy5 was incorporated. These particles were labelled with ^{124}I and surface functionalized with peptide cyclo-(Arg-Gly-Asp-Tyr) (cRGDY) to selectively target integrin-expressing tumours. C dots were synthesized in such a way that they had limited reticuloendothelial system (RES) uptake and promote renal excretion. Preliminary experiments on in vivo safety revealed that these fluorescent silica nanoparticles were safe and did not show any toxicity in mice. These particles were also found to be an effective bioimaging probe for cancer imaging. Burns et al. carried out in vivo biodistribution studies of the developed C dots in nude mice wherein nanoparticles were injected intravenously. The particles were found to show rapid renal clearance within 45 min of injection and majority of these particles accumulated in the liver. To further modify the clearance, the particles were coated with methoxy-terminated poly(ethylene glycol) chains. By careful manipulation of the surface features of C dots, they can be used for wide variety of biomedical applications including imaging and therapy. Based on the encouraging results of pre-clinical studies, these nanoparticles received approval from the FDA as an Investigational New Drug (IND) to conduct the human clinical trial, phase I. The first human clinical trials suggested its safety for human use. A pilot clinical trial was conducted in five metastatic melanoma patients to assess the pharmacokinetic (PK) profile of C dots following a single injection dose. The PK profiles, renal excretion, metabolic profile assessment in patients suggested that the particles were well tolerated, preferentially accumulated in the tumour site and were found to be safe for human use.

A study wherein the potential of ordered mesoporous silica nanoparticles (OMS) in enhancing the bioavailability of fenofibrate in man was conducted by Bukara and collaborators which can be considered as another breakthrough step acting as a trigger in evoking interest among the researchers for the use of MSNs for biomedical applications. Promising results obtained by them in their preclinical studies prompted them to complement those results with clinical studies. Fenofibrate was loaded into OMS and these were subsequently enclosed within capsules. The study was carried out with 12 volunteers who were administered a single dose of fenofibrate OMS and the marketed formulation of fenofibrate, Lipanthyl. Safety assessment was performed by periodic monitoring of the vital signs, 12-lead electrocardiogram (ECG) and blood biochemical parameters in the subjects. The PK profile revealed an increase in the rate and extent of absorption of fenofibrate when incorporated in OMS as compared to the marketed product. In addition, the formulation was found to be well tolerated in the volunteers ensuring the safety of the developed OMS formulation.

Aerogel

Aerogels are a fascinating class of nanoporous materials derived by sol-gel technique which possess sophisticated potentialities for a range of applications. They possess unique properties such as high porosity, large inner surface area, low density, small index of refraction, low sound velocity, low thermal conductivity and low dielectric constant. Aerogels are quite transparent since they scatter light weakly because of the extremely small particle and pore sizes. Important structural properties of aerogels are provided in table.

Table: Properties of silica aerogel.

Property	Range
Bulk density (g/cc)	0.003-0.500
Skeletal density (g/cc)	1.700-2.100
Porosity (%)	80-99.8
Mean pore diameter(nm)	20-150
Inner surface area (m^2 /g)	100-1600
Refractive index	1.007-1.24
Thermal conductivity (in air 300)(w/(mk)	0.017-0.021
Modulus of elasticity (mpa)	0.002-100
Sound velocity cl (m/s)	<20-800
Dielectric constant	1-2 (18-40 GHz)

Aerogels find applications as pesticides, in cosmic dust capture, waste water treatment, moulds for casting aluminium metal, aerocapacitors, absorbing media for dessication, heat storage device for automobiles, encapsulation media, hydrogen fuel storage material, ideal dielectric for ultra fast integrated circuits, thermal insulators, fillers, particle detectors, catalysts, gas sensors, light guides, capacitors. Synthesis of aerogel starts with the controlled conversion of a sol into a gel. The growth of clusters or polymer chains from a chemical solution occurs initially and these primary entities cross-link leading to the formation of a coherent network-still embedded in a liquid. The structure of aerogel is characterized by well accessible, cylindrical branched mesopores. Nitrogen adsorption studies, mercury porosimetry, microscopic techniques and scattering techniques (SAXS and SANS) are used to explore the structure of aerogels. Schematic diagram of aerogel structure. When a silica aerogel monolith breaks under force only secondary particle lose contact with one another while primary particle remains undisturbed. So the mesoporous space between these particles would be maintained. The silica aerogels exhibit microstructures consisting of loosely interconnected silica particles which form an open network.

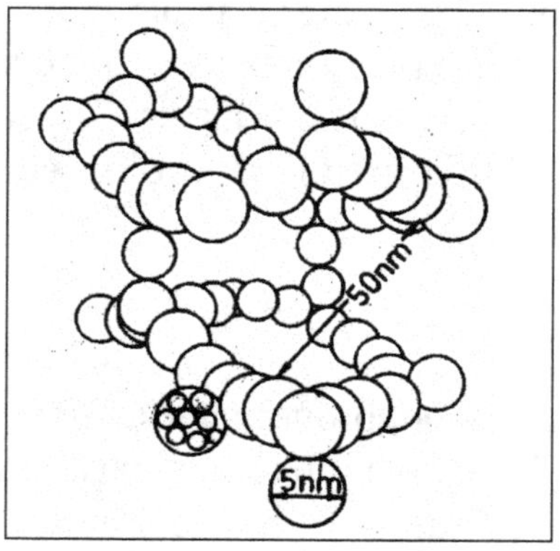

Structure of aerogel.

The first aerogels were made by S. S. Kistler in 1931 from a water glass based recipe. Later Teichner et al. improved this sol-gel process. They hydrolysed TMOS by adding controlled amount of water in a methanol medium. Silica form a coherent network embedded in alcohol. This alcogel was then successfully dried under hypercritical conditions (TC=239 °C, PC=81 bar). Using a two step sol-gel process and CO_2 drying, a group at Lawrence Livermore National Laboratory (LLNL) prepared ultra low density aerogels with densities as low as 3 Kg/m^3 . The potential of the material was only widely recognized in the 1970's when the aerospace industry was looking at the possibility of developing extremely light thermal insulator. Further developments resulted in using aerogel as insulator in pathfinder mission to mars, for insulating the sensitive electronics from harsh temperature of mars. Conventional method of synthesis of aerogel is by a sol-gel process followed by supercritical drying of the wet gel in an autoclave. A schematic diagram which shows the process of synthesis of aerogel is provided as figure. During initial stages of drying, the capillary forces caused by the evaporation of solvents from pores in the gel, create an overall drying stress and local differential stresses owing to non uniform pore size distribution.45 Silica aerogels are ultra pure gels derived by hydrolysis-condensation reaction of silica alkoxide or salts.

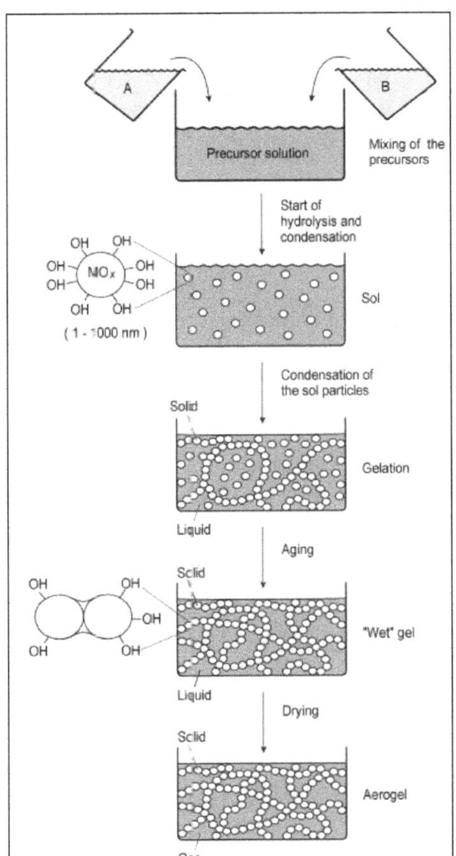

General scheme for the synthesis of aerogels by sol-gel processing.

Supercritical Drying

Supercritical drying involves the process of drying a wet gel in an autoclave above the critical temperature and pressure of its solvent. At the critical point of drying there is no longer any distinction between the liquid and vapour phases, and there is no liquid-vapour interface or capillary pressure. The resulting gels called aerogels have about 99% porosity, extremely low densities and a high inner surface area. A schematic representation of supercritical drying is provided in figure. At the critical point [Tc, pc] the density of the liquid and the gas are equal. Supercritical drying can be performed along path A or B. However this technique is expensive and hazardous, and results in fragile aerogels having low strength.

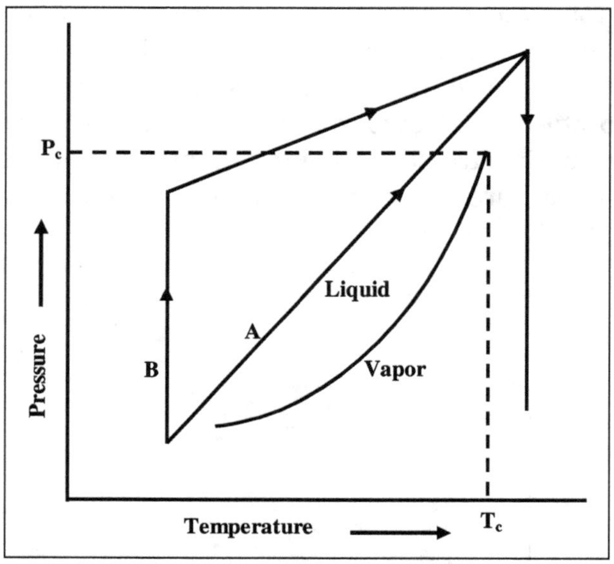

Schematic representation of supercritical drying.

Subcritical Drying

Subcritical drying technique, a less expensive method for the synthesis of aerogel, which involves controlled solvent exchange, aging and surface modification is evolving as a promising technique for the synthesis of aerogel at ambient temperature and pressure.

Solvent	Tc (°C)	Pc (MPa)
Methanol	240	7.9
Ethanol	243	6.3
Acetone	235	4.7
Isopropanol	235	4.7
H_2O	374	22.1
CO_2	31	7.3

Ambient pressure methods for silica aerogels include surface modification, network strengthening and use of DCCAs. The basic concept of this technique is to improve the strength and stiffness of the gel network and avoid shrinkage or crack during drying. The extent of shrinkage during drying is governed by the capillary pressure exerted by the pore fluid and the modulus of the solid matrix. Solvent exchange with a low surface tension liquid can reduce the capillary pressure.

References

- Soil Science Glossary Terms Committee (2008). Glossary of Soil Science Terms 2008. Madison, WI: Soil Science Society of America. ISBN 978-0-89118-851-3
- Nanoporous-materials-27264580: researchgate.net, Retrieved 19 July, 2020
- Eftekhari, Ali (2017). "Ordered Mesoporous Materials for Lithium-Ion Batteries". Microporous and Mesoporous Materials. 243: 355–369. doi:10.1016/j.micromeso.2017.02.055
- Soil Science Glossary Terms Committee (2008). Glossary of Soil Science Terms 2008. Madison, WI: Soil Science Society of America. ISBN 978-0-89118-851-3
- Porosity-process-technologies-and-applications/nanoporous-carbon-synthesis-an-old-story-with-exciting-new-chapters: intechopen.com, Retrieved 26 March, 2020
- Nelson B (2018). "Add another strange property to the list of silver's bizarre traits". MNN. Narrative Content Group. Retrieved 9 Feb 2019

Metal Matrix Nanocomposites

5

- Lightweight Metal Matrix Nanocomposites
- Carbon Nanotube Metal Matrix Composites
- Electrodeposition of Metal Matrix Nanocomposites

The nanocomposites which consist of metal as the matrix, and ceramic or organic compounds as the reinforcement are called metal matrix nanocomposites. These can be classified into continuous and non-continuous reinforced materials. This chapter delves into the subject of metal matrix nanocomposites for an in-depth understanding of the subject.

Metal-matrix composites (MMCs) possess excellent physical and mechanical properties. The filler reinforcement into the metallic matrix improves the stiffness, specific strength, wear, creep and fatigue properties compared to the conventional engineering materials.

Synthesis of Metal Oxide Nanocomposites (MONC)

The synthesis of uniform sized nanocomposite is very important because their properties include optical, magnetic, electrical and biological properties depending on their size and dimensions. The synthetic methods are frequently classified in to three classes i.e. solution based synthesis, vapor phase synthesis and gas phase synthesis. Another approach is to divide these synthetic approaches into two broad categories i.e. a) top-down approach which includes physical methods and b) bottom up approach which encompasses wet methods. The choice of synthetic approach depends on the desired characteristics required in the nanoparticles composites e.g. size, morphology, crystal structure etc. but the benefit of physical methods is the production of large amount of nanocomposites but synthesis of equal sized nanocomposite particles is not easily attainable. In comparison, wet chemical methods give uniformity in size of nanocomposites as controlled particle size can be achieved easily. Although by varying conditions of reaction, different shapes (nano-rods, nanowires, nanotubes etc.) of nanocomposites can also be synthesized.

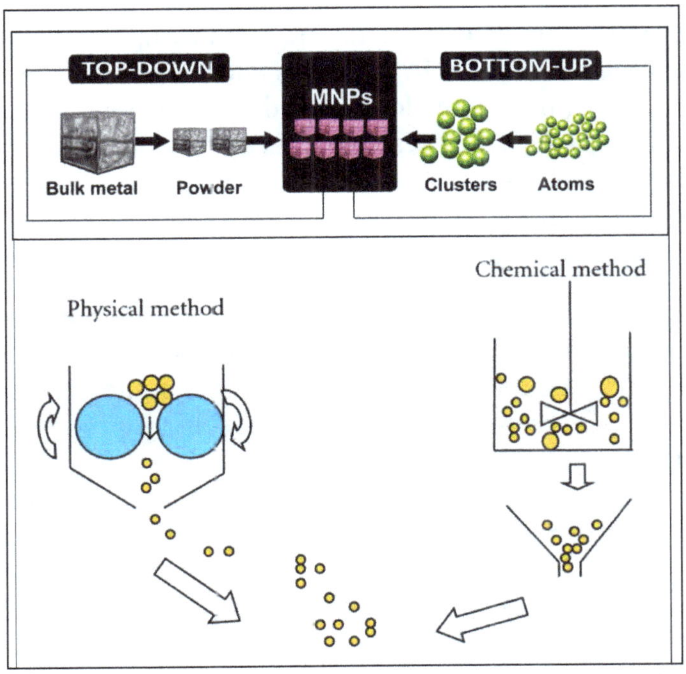

Schematic representations of top down and bottom up approaches.

There are different wet methods for synthesis of metal nanoparticles and nanocomposites but coprecipitation, sol gel and hydrothermal methods are cost effective as compared to other methods so these are discussed in detail.

Co-precipitation Method

The co-precipitation method is used for the synthesis of metal oxides nanoparticles, mixed metal or metal ceramics nano-composites, produces precipitate that are separated from solution. Inorganic salts are used as precursors, dissolved in water and other solvents to obtain homogenous solution of ions, and then these salts start precipitating as hydroxides or oxalates when the critical concentration of species is attained followed by nucleation and growth phases. The size and shape of particles is greatly influenced by solution pH, temperature and concentration of salt. After precipitation, filtration and washing is done followed by calcination to convert hydroxide into oxides with a definite crystalline structure.

The precipitating medium usually employed includes NaOH, NH_3 or NH_4OH, Na_2CO_3 etc. The use of surfactants is also a common practice to avoid agglomeration which also tampers the particle size of the composites obtained by this technique.

The method offers the advantage of being low cost, simple, water based reaction, flexibility, mild reaction conditions and size control.

Sol-gel Method

Sol-gel method gained attention as a promising method for the synthesis of nanomaterials owing to their mild reaction conditions and building up the materials from molecular precursors leading to variation in materials and properties.The resulting product of sol-gel method is either films or colloidal powder. The sol-gel method has capability of producing micro and nanostructures. The size, shape and structure of final product are greatly influenced by the reaction parameters. The process involves the simple wet chemical reaction based on hydrolysis and condensation leading to formation of sol which through the process of aging results in formation of an integrated network as gel.

The sol–gel method is also very attractive for the synthesis of nanostructures containing more than one component, since the slow reaction kinetics allows good structural engineering of the final product. Another advantage is that the reactions are conducted at low temperatures or at room temperature. The sol–gel process involves inorganic precursors that undergo various chemical reactions, resulting in the formation of a three-dimensional molecular network. One of the most common routes is via hydrolysis and condensation of metal alkoxides to form larger metal oxide molecules that polymerize to form the coating. The sol–gel procedure allows coating of substrates with complex shapes on the nanometer to micrometer scale, which some commonly used coating procedures cannot achieve. The substrates include colloidal particles, organic/ inorganic crystals, or even fibers and nanotubes.

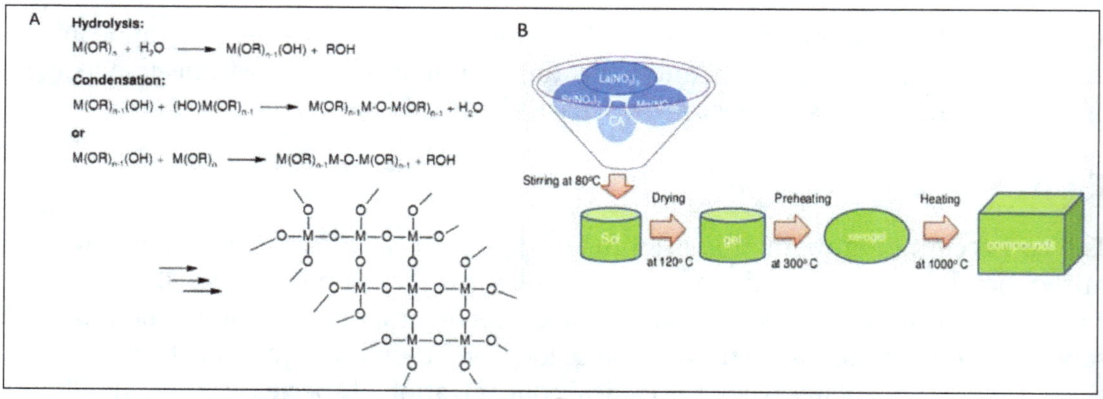

A) Chemical steps and B) Schematic representation of Sol-gel method.

Hydrothermal Method

The hydrothermal method involves the heterogeneous chemical reaction in a solvent (aqueous or non-aqueous) occurring above room temperature and at pressure more than 1atm in a closed system. To modify the size and properties the use of surfactants, capping agents, mineralizers is a common practice. The new trend is to use this technique in combination with microwave, sol-gel that can not only vary the physiochemical and structural properties of the materials but in addition to that can result in formation of single phased materials with enhanced stability. Further just by altering the temperature, time and pressure of the reaction particle size, phase changes morphology and properties as presented in figure.

Effect of time and temperature on size and morphology of $ZnO-SnO_2$ nanocomposite A) 160 °C for 30min B) 160 °C for 60min C) 160 °C for 4h D) 160 °C for 12h E) 180 °C for 30min F) 200 °C for 12h.

Characterization of Metal Oxide Nanocomposites (MONC)

The nanocomposites are characterized using different techniques to get insight into the morphology, particle size, phase, composition, thermal stability, optical, magnetic, electrical and thermal properties.

X-ray Diffraction (XRD)

Powder X-ray Diffraction (XRD) is used for phase determination and unit cell information of the nanocomposites under investigation. The technique is also frequently employed to determine particle size using Scherrer's formula.

$$D = K\lambda / \beta \cos\theta$$

where D (nm) is the mean size of the crystalline domains, K is the dimensionless shape factor which value is close to unity, λ is x-ray wavelength and β is line broadening referring to FWHM and θ is Bragg angle. The formula is applicable when nanocomposites have definite crystalline structure while in case of amorphous structure this formula is not applicable for particle size.

In majority of cases the phase determination is the usual information obtained from PXRD technique especially when comparing the synthetic approach or effect and present of doping in the matrix structure. Kalantariet.al. used PXRD as a means to confirm surface binding of Fe^{3+} ions with talc used as matrix in nanocomposite by shifting in d-spacing value of the talc while constant 2θ and FWHM values in matrix and MONC were taken as grantee for talc maintaining its structure. Further the technique is also helpful in determining the nanoscale dispersion in polymer matrix when used in combination with TEM but when used independently the low loading in the matrix structure might lead to negative results as the diffraction peaks are usually not observed. The exfoliation of graphite oxide (GO) by inclusion of metal oxides as indicated by decrease in intensity of GO peak in the nanocomposites was also determined by various researches. The researchers had also used the technique for determining the presence and absence of impurities from matrix structure by the appearance of peaks in the matrix as shown in figure in which the disappearance of peaks in purified spectra of MWCNT after acid washing confirms the removal of impurities.

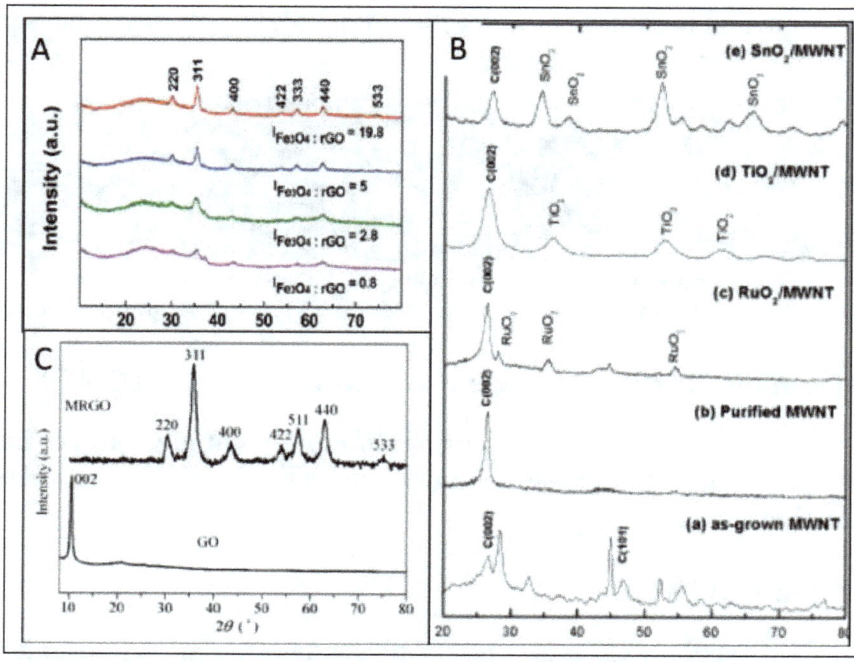

Powder XRD patterns of A) Fe_3O_4/GO at different ratios, B) MO/MWCNT and C) MnO_2/GO indicating disappearance of GO peak.

Microscopic Techniques - SEM/AFM/TEM

The microscopic techniques focused in the chapter are scanning electron microscopy (SEM), atomic force microscopy (AFM) and transmission electron microscopy (TEM). SEM is mostly employed for studying the surface morphology of the MONC but that requires the surface to be electrically conductive hence in case of non-conducting samples the thin layer of gold or carbon is coated. The resolution of SEM is around 1-2 nm. The effect of doping of different metal oxide nanoparticles on the matrix, agglomeration, shape etc.

TEM technique has gained much attention as compared to SEM because of its better resolution i.e. 0.1-0.2nm and wealth of information that can be extracted from it and especially from high resolution TEM (HRTEM). The core-shell structure of the nanocomposites, doping effects on morphology, layering in structure, particle size, gelling agents impact, surface roughness, nanoparticle dispersion etc. can be determined. In this regard use of dark field and bright field images can help in getting more information about the structural changes. The main limitation of the technique is in sample preparation which requires low film thickness.

AFM enables surface imaging of both conducting and non-conducting materials at atomic resolution level by measuring interactive forces between the atom of sample and AFM tip. It can work under different modes i.e. contact mode, tapping mode and conductive mode. 2D, 3D and line profile data obtained from AFM can give information about the dispersion of the nanoparticles, height of nanoparticles dispersed in matrix and height of matrix itself. The technique also helps in determining the surface roughness of the nanocomposite membrane.

SEM images showing effect of doping on (A) purified MWCNT matrix with (B) RuO_2, (C) TiO_2 and (D) SnO_2, respectively; (E, F) effect of varying carbon to metal ratio on agglomeration of CNT-Fe-Al_2O_3 i.e. C=4.8% and 5.7% respectively; and (G, H) effect of doping of Fe_3O_4 and CuO on PANi, respectively.

Thermal Stability

Thermal analysis including thermogravimetric analysis (TGA), differential scanning calorimetry (DSC), dynamic mechanical thermal analysis (DMTA), thermal mechanical analysis (TMA) etc. is a useful tool to determine the stability of the nanocomposite materials and various phenomena occurring because of doping, curing and annealing. The steps leading to thermal decomposition of the nanocomposites, endothermal or exothermic nature of decomposition process, solvent/moisture loss, weight loss at each step, final decomposed matter etc. are the usual information obtained by the TA. The content of MONP present in the matrix was also assessed by the residue left after complete thermal decomposition of the polymer matrix achieved. Laachachi et.al. compared the thermal stability of the PMMA by doping equal amount of TiO_2 and Fe_2O_3 and observed with help of onset degradation temperature that former imparts better stability than the later. He also used the technique to determine the best wt.% of metal oxide NPs in the polymer matrix that causes decrease in heat release – the property considered good for flame retardants. The change incuring initiating temperature of the resin with introduction of functionalized and non-functionalized nanoparticles and treatments was also tied by changes in TGA spectra. TGA of MnO_2/CNT MONC helped in determining the interlayer water from which nanocomposite formula was derived that further elucidated the oxidation state of Mn in the NC.

Applications of Metal Oxide Nanocomposites (MONC)

The applications of metal oxide nanocomposites in different sectors had been widely recognized. The chapter will focus on the four main categories i.e. environment, agriculture/ food and health/ medicine.

Application in Environment

The MONC are frequently used as adsorbents, photocatalysts and sensors to tackle environmental pollution problems. The advantage of being at nanoscale offers the high surface area and high reactivity making them highly effective for water purification and early sensing of pollutants. Metal oxides nanoparticles are used in combination with graphene, silica, other oxides, CNT, polymers for the removal of dyes, heavy metals etc.

Application in Agriculture and Food

The MONC are widely used in packaging of foodstuff, not only provides strength as filler material (silicates, clays, TiO_2) but also as antimicrobial agent like silver oxide nanocomposites and oxygen scavengers. While in agriculture sector, MONC used as nanosensor for pesticide and pathogen detection in plants and source for delivery of genetic material for improvement of crops. Metal oxide nanoparticles (ZnO, CeO and CuO) and their nanocomposite (with fertilizers and zeolite) used as slow and controlled release of fertilizers provide nutrients to plants for prolonged period and also helps in prevent of soil degradation and improvement of sustainable agriculture.

Application in Health and Medicine

The MONC have many applications in medicine, for delivery of drugs and imaging, diagnosis and screening of diseases, DNA sequencing, in gene therapy and tissue culturing and in cancer treatment.

Metal-matrix Composites for Space Applications

The extreme environment in space presents both a challenge and opportunity for material scientists. In the near-earth orbit, typical spacecraft encounter naturally occurring phenomena such as vacuum, thermal radiation, atomic oxygen, ionizing radiation, and plasma, along with factors such as micrometeoroids and human-made debris. For example, the International Space Station, during its 30-year life, will undergo about 175,000 thermal cycles from +125 °C to −125 °C as it moves in and out of the Earth's shadow. Re-entry vehicles for Earth and Mars missions may encounter temperatures that exceed 1,500 °C. Critical spacecraft missions, therefore, demand lightweight space structures with high pointing accuracy and dimensional stability in the presence of dynamic and thermal disturbances. Composite materials, with their high specific stiffness and low coefficient of thermal expansion (CTE), provide the necessary characteristics to produce lightweight and dimensionally stable structures. Therefore, both organic-matrix and metal-matrix composites (MMCs) have been developed for space applications.

Despite the successful production of MMCs such as continuous-fiber reinforced boron/ aluminum (B/Al), graphite/ aluminum (Gr/Al), and graphite/ magnesium (Gr/Mg), the technology insertion was limited by the concerns related to ease of manufacturing and inspection, scale-up, and cost. Organic-matrix composites continued to successfully address the system-level concerns related to microcracking during thermal cycling and radiation exposure, and electromagnetic interference (EMI) shielding; MMCs are inherently resistant to those factors. Concurrently, discontinuously reinforced MMCs such as silicon-carbide particulate (p) reinforced aluminum (SiC_p/Al) and G_p/Al composites were developed cost effectively both for aerospace applications (e.g., electronic packaging) and commercial applications.

Historically, MMCs, such as steel-wire reinforced copper, were among the first continuous-fiber reinforced composites studied as a model system. Initial work in late 1960s was stimulated by the high-performance needs of the aerospace industry. In these development efforts, performance, not cost, was the primary driver. Boron filament, the first high-strength, high-modulus reinforcement, was developed both for metal- and organic- matrix composites. Because of the fiber-strength degradation and poor wettability in molten-aluminum alloys, the early carbon fibers could only be properly reinforced in organic-matrix composites. Therefore, the development of MMCs was primarily directed toward diffusion-bonding processing. At the same time, optimum (air stable) surface coatings were developed for boron and graphite fibers to facilitate wetting and inhibit reaction with aluminum or magnesium alloys during processing.

METAL-matrix Composite Processing

Three processing methods have been primarily used to develop MMCs: High-pressure diffusion bonding, casting, and powder-metallurgy techniques. More specifically, the diffusion-bonding and casting methods have been used for continuous- fiber reinforced MMCs. Discontinuously reinforced MMCs have been produced by powder metallurgy and pressure-assist casting processes. MMCs such as B/Al, Gr/Al, Gr/Mg, and Gr/ Cu have been manufactured by diffusion bonding for prototype spacecraft components such as tubes, plates, and panels.

Properties

Table lists the typical properties of a few continuous-fiber reinforced MMCs. Generally, measured properties of as-fabricated MMCs are consistent with the analytically predicted properties of each composite. The primary advantage of MMCs over counterpart organic-matrix composites is the maximum operating temperature. For example, B/Al offers useful mechanical properties up to 510 °C, whereas an equivalent B/Ep composite is limited to about 190 °C. In addition, MMCs such as Gr/Al, Gr/Mg, and Gr/Cu exhibit higher thermal conductivity because of the significant contribution from the metallic matrix.

Table: Material properties of unidrectional metal-matrix composites for space applications.

Properties	P100/6061 Al (0°)	P100/AZ91C Mg (0°)	Boron/Al (0°)
Volume Percent Reinforcement	42.2	43	50
Density, r (gm/cm³)	2.5	1.97	2.7
Poisson Ratio n_{xy}	0.295	0.3	0.23
Specific Heat C_p (J/kg-K)	812	795	801
Longitudinal			
Young's Modulus (x) (GPa)	342.5	323.8	235
Ultimate Tensile Strength (x) (MPa)	905	710.0	1100
Thermal Conductivity K_x (W/m-K)	320.0	189	—
CTE_x (10^{-6} /K*)	-0.49	0.54	5.8
Transverse			
Young's Modulus (y) (GPa)	35.4	20.7	138
Ultimate Tensile Strength (y) (MPa)	25.0	22.0	110
Thermal Conductivity K_y (W/m-K)	72.0	32.0	—

Table below lists the properties of discontinuously reinforced aluminum (DRA) composites for spacecraft and commercial applications. DRA is an isotropic MMC with specific mechanical properties superior to conventional aerospace materials. For example, DWA Aluminum Composites has produced MMCs using 6092 and 2009 matrix alloys for the best combination of strength, ductility, and fracture toughness, and 6063 matrix alloy to obtain high thermal conductivity. Similarly, Metal-matrix Cast Composite (MMCC) Inc. has produced graphite particulate-reinforced aluminum composites for the optimum combination of high specific thermal conductivity and CTE.

Table: Material properties of discontinuously reinforced aluminum-matrix composites.

Properties	Graphite Al GA 7-230	Al6092/SiC/17.5p	Al/SiC/63p
Density, r (gm/cm³)	2.45	2.8	3.01
Young's Modulus (GPa)	88.7	100	220
Compressive Yield Strength (MPa)	109.6	406.5	
Tensile Ultimate Strength (MPa)	76.8	461.6	253
Compressive Ultimate Strength (MPa)	202.6	—	
CTE (x-y) (10^{-6} /K)	6.5-9.5	16.4	7.9
Thermal Conductivity (W/m-K) (x-y)	190	165	175
(z)	150		170
Electrical Resistivity (μ-ohm-cm)	6.89	—	

Application

While the desire for high-precision, dimensionally stable spacecraft structures has driven the development of MMCs, applications thus far have been limited by difficult fabrication processes. The first successful application of continuous-fiber reinforced MMC has been the application of B/Al tubular struts used as the frame and rib truss members in the mid-fuselage section, and as the landing gear drag link of the Space Shuttle Orbiter. Several hundred B/Al tube assemblies with titanium collars and end fittings were produced for each shuttle orbiter. In this application, the B/Al tubes provided 45% weight savings over the baseline aluminum design.

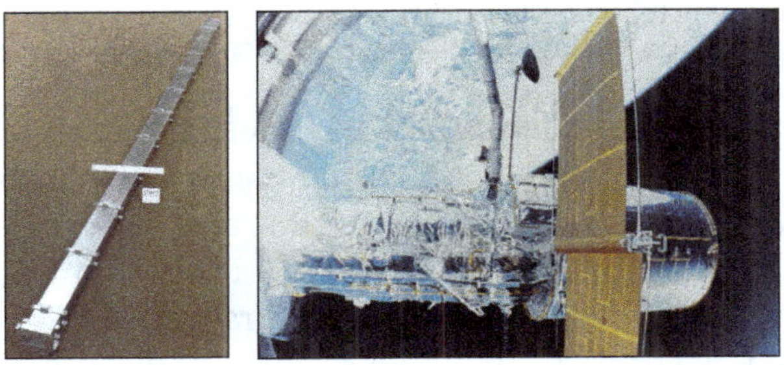

The P100/6061 Al high-gain antenna wave guides/ boom for the Hubble Space Telescope (HST) shown (a-left) before integration in the HST, and (b-right) on the HST as it is deployed in low-earth orbit from the space shuttle orbiter.

The major application of Gr/Al composite is a high-gain antenna boom for the Hubble Space Telescope made with diffusion-bonded sheet of P100 graphite fibers in 6061 Al. This boom (3.6 m long) offers the desired stiffness and low CTE to maintain the position of the antenna during space maneuvers. In addition, it provides the wave-guide function, with the MMC's excellent electrical conductivity enabling electrical-signal transmission between the spacecraft and the antenna dish. Also contributing to its success in this function is the MMC's high dimensional stability—the material maintains internal dimensional tolerance of ±0.15 mm along the entire length. While the part currently in service is continuously reinforced with graphite fibers, replacement structures produced with less expensive DRA have been certified.

Mid-fuselage structure of Space Shuttle Orbiter showing boron-aluminum tubes.

P100/AZ91C Gr/Mg tubes produced by the vacuum-assist casting process: (a-left) as-cast tubes, and (b-right) demonstration Gr/Mg truss structure.

Like the Gr/Al structural boom, a few MMCs have been designed to serve multiple purposes, such as structural, electrical, and thermal-control functions. For example, prototype Gr/Al composites were developed as structural radiators to perform structural, thermal, and EMI-shielding functions. Also, Gr/Cu MMCs with high thermal conductivity were developed for high-temperature structural radiators. A DRA panel is used as a heat sink between two printed circuit boards to provide both thermal management and protection against flexure and vibration, which could lead to premature failure of the components in the circuit board.

Cast SiCp/Al attachment fittings: (a-top) multi-inlet fitting for a truss node, and (b-bottom) cast fitting brazed to a Gr/Al tube.

In technology-development programs sponsored by the U.S. Defense Advanced Research Projects Agency and the U.S. Air Force, graphite/magnesium tubes for truss-structure applications have been successfully produced (jointly by Lockheed Martin Space Systems of Colorado and Fiber Materials of Maine) by the filament-winding vacuum-assisted casting process. Figures show a few of the cast Gr/Mg tubes (50 mm dia × 1.2 m long) that were produced to demonstrate the reproducibility and reliability of the fabrication method.

Of the DRA composites, reinforcements of both particulate SiC_p/Al and whisker (w) SiC_w/Al were extensively characterized and evaluated during the 1980s. Potential applications included joints and attachment fittings for truss structures, longerons, electronic packages, thermal planes, mechanism housings, and bushings. Figures 4a and 4b show a multi-inlet SiC_p/Al truss node produced by a near net-shape casting process.

Because of their combination of high thermal conductivity, tailorable CTE (to match the CTE of electronic materials such as gallium arsenide or alumina), and low density, DRA composites are especially advantageous for electronic packaging and thermal-management applications. Several SiC_p/Al and Gr_p/Al electronic packages have been space-qualified and are now flown on communication satellites and Global Positioning System satellites. These components are not only significantly lighter than those produced from previous metal alloys, but they provide significant cost savings through net-shape manufacturing. DRA is also used for thermal management of spacecraft power semiconductor modules in geosynchronous earth-orbit communication satellites, displacing Cu/W alloys with a much higher density and lower thermal conductivity, while generating a weight savings of more than 80%. These modules are also used in a number of land-based systems, which accounts for an annual production near 1 million piece-parts. With these demonstrated benefits, application of DRA MMCs for electronic packages will continue to flourish for space applications.

Lightweight Metal Matrix Nanocomposites

Composite materials that traditionally incorporate micron scale reinforcements in a bulk matrix offer opportunities to tailor material properties such as hardness, tensile strength, ductility, density, thermal and electrical conductivity, and wear resistance. With the advent of nanomaterials, nanocomposites are envisioned, and are being developed, with properties that overcome the limitations for metals or composites that contain micron scale reinforcements. For example, carbon nanotubes have been shown to exhibit ultra-high strength and modulus, and have anisotropic electrical conductivity. When included in a matrix, carbon nanotubes could impart significant property improvements to the resulting nanocomposite.

In the past decade, much work has gone towards developing polymer matrix nanocomposites and many such materials are already used in various applications. Metallic composites containing nanoparticles or carbon nanotubes could offer distinct advantages over polymeric composites due to the inherent high temperature stability, high strength, high modulus, wear resistance, and thermal and electrical conductivity of the Metal-matrix Aluminum nanocomposites are predicted to surpass the weight reduction currently realized through the use of polymer-based nanocomposites and polymer-based fiber composites in aerospace applications primarily because these metal matrices have higher strength and stiffness. They also have much better thermal stability.

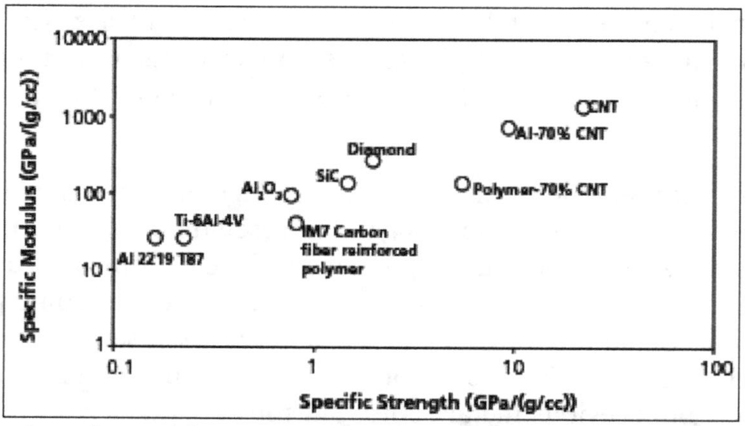

Comparison of potential materials and reinforcements for aerospace applications.

The development of Metal-matrix Nanocomposites (MMNCs), however, is still in its infancy. The MMNCs synthesized to date include Al-B_4C, Mg-SiC, Al-CNT, Cu-CNT and Ti-SiC, prepared using powder metallurgy, and Al-SiC, Mg-SiC, Al-Al_2O_3, Al- CNT, Mg-Y_2O_3, Al-Diamond, and Zn-SiC, prepared using solidification processing.

Though there is great potential for the use of MMNCs in a variety of applications, their use is hindered by their cost, difficulty in the manufacture of large complex shaped parts and their often poor ductility.

Synthesis, Processing and Properties of Metal-matrix Nanocomposites

The greatest challenges facing the development of MMNCs for wide application are the cost of nanoscale reinforcements and the cost and complexity of synthesis and processing of nanocomposites using current methods. As with conventional Metal-matrix composites with micron-scale reinforcements, the mechanical properties of a MMNC are strongly dependent on the properties of reinforcements, distribution, and volume fraction of the reinforcement, as well as the interfacial strength between the reinforcement and the matrix. Due to their high surface area, nanosize powders and nanotubes will naturally tend to agglomerate to reduce their overall surface energy, making it difficult to obtain a uniform dispersion by most conventional processing methods. In addition, due to their high surface area and surface dominant characteristics, these materials may also be highly reactive in metal matrices. For example, in Al/CNT composites there are concerns that brittle aluminum carbide phases could form during processing, impairing the mechanical properties and electrical conductivity of the nanocomposite. Because of these concerns, processing methods are being developed to produce MMNCs with uniform dispersions of nanomaterials and little deleterious interfacial reactions.

The methods that have been used to synthesize Metal-matrix nanocomposites include powder metallurgy, deformation processing, vapor phase processing, and in some cases solidification processing. Powder metallurgy involves the preparation of blends of powders of metal and reinforcements, followed by consolidation and sintering of the mixtures of powders to form the part. Deformation processing involves subjecting a metal to high rates of deformation to create nanostructured grains in a Metal-matrix Vapor phase processing methods such as chemical vapor deposition (CVD) can be used to deposit thin films creating dispersed multiphase microstructures, multilayered microstructures, or homogeneous nanostructured coatings. Each of these methods can create very desirable microstructures, however they are expensive and difficult to scale up to manufacture large and complex shapes in bulk.

Of the processing methods available for synthesis and processing of MMNCs, the least expensive method for production of materials in bulk is solidification processing. There are various avenues by which researchers have created nanostructures and nanocomposite materials using solidification and these can be divided into three categories:

- Rapid solidification.
- Mixing of nanosize reinforcements in the liquid followed by solidification.
- Infiltration of liquid into a preform of reinforcement followed by solidification.

Rapid solidification (implying solidification rates of up to 10^4–10^7 °C/s) through methods such as melt spinning (a liquid metal stream is impinged on a spinning copper drum), or spray atomization (a superheated liquid metal is atomized with gas jets and impinged on a substrate) can lead to nanosize grains as well as amorphous metals from which nanosize reinforcements can be precipitated in the amorphous matrix during heating to form nanocomposites. Mixing techniques involve adding particulate reinforcements and mechanically dispersing them in the matrix. Mixing methods that have been applied to synthesize MMNCs include stir mixing, where a high temperature impeller is used to stir a melt that contains reinforcements, creating a vortex in the melt, and ultrasonic mixing, where an ultrasonic horn is used to create cavitation in the melt that disperses the particulate reinforcements by a gas streaming effect that occurs through the collapse of bubbles within the melt. Infiltration techniques entail infiltrating a preform or partial matrix containing the reinforcements with a liquid metal. The preform consists of particles formed in a particular shape with some binding agent, and can be composed of the additives and binding agent alone or with some portion of the matrix added as a partial filler. Infiltration methods that have been used include ultra high pressure, where the pressure used to infiltrate a high-density preform of nanoparticles is in excess of 1 Gpa, and pressureless infiltration, where a block of metal is melted on top of a lower density preform of nanoparticles and allowed to seep into the preform.

Figure illustrates examples of the different microstructures of Aluminum Alloy-Al_2O_3 nanoparticle composites that have been produced by different processing methods. Figure shows a microstructure exhibited by a cast nanocomposite synthesized at

UW–Milwaukee. This MMNC was made using a unique casting method combining the use of stir mixing, and ultrasonic mixing, with a wetting agent added to the molten alloy to incorporate nanoparticles in a Metal-matrix This process resulted in the incorporation of nanoparticles within microscale grains of aluminum, and formed a bimodal microstructure. Ceramic nanoparticles may be uniformly dispersed in metal matrices to increase the tensile strength and wear resistance, using methods such as ultrasonic cavitation of the melt to further disperse the particles. The influence of nanoscale reinforcements on formation of solidification microstructure including their influence on nucleation, growth, particle pushing, solute redistribution, heat and fluid flow, however, will have to be understood to reproducibly create desired structures. While there is some understanding of the influence of micron size particles on the formation of solidification microstructures, the influence of nanoscale reinforcements on each of the constituents of solidification structure formation need to be studied using both theoretical and experimental research. Figure shows a transmission electron microscope (TEM) micrograph of the microstructure obtained by ball milling pure metals and nanopowders, followed by hot pressing/sintering to form a nanocomposite. The matrix aluminum and the Al_2O_3 powders are nanosize in this TEM micrograph and the reinforcement phases are mainly restricted to the grain boundaries, which will likely have a grain boundary pinning effect. Since a much higher percentage of Al_2O_3 has been added using this process, the resulting sample exhibited a substantial improvement in its wear properties.

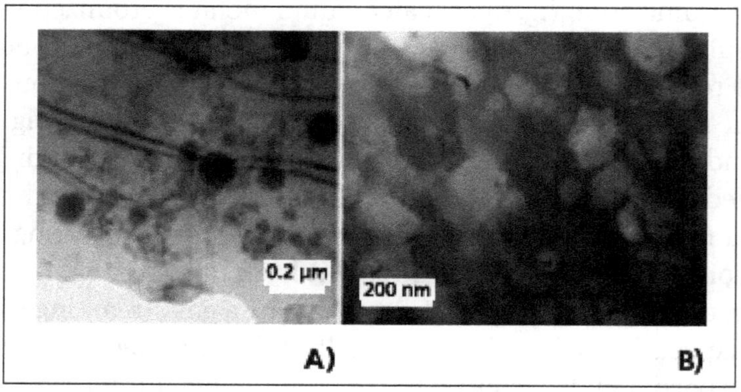

TEM of Aluminum Alloy-Al_2O_3 nanocomposites produced by liquid and solid based methods. A) Stir cast A206- 2 vol % Al_2O_3 (47 nm) nanocomposite produced at UW–Milwaukee. B) Powder metallurgy based Aluminum alloy-15 vol% Al_2O_3.

Great improvements can be achieved in specific material properties by adding only a small percentage of a dispersed nano-phase as reinforcement. Table presents selected studies on MMNCs, the processing techniques used and their respective properties. When the reinforcement in a Metal-matrix is brought down from micron-scale to nano-scale, the mechanical properties are often substantially improved over what could be achieved using micron-scale reinforcements. This is possibly due to the exceptional properties of the individual nano-phase reinforcements themselves, smaller means free path between neighboring nanoparticles and the greater constraint provided by the higher surface area of nano-phase reinforcements. Nano-phase reinforcements like CNT and SiC have much higher strengths than similar micron-scale reinforcements. In some cases, the nanoscale reinforcement leads to property changes in the matrix itself. For instance, nanoscale reinforcements can lead to nanosize grains in the matrix, which will increase the strength of the matrix. Due to their size, properties of nanomaterials are dominated by their surface characteristics, rather than their bulk properties, which is the case in micronscale reinforcements. The potentially unique interfaces between nanosized reinforcements and the matrix can lead to even greater improvements in the mechanical properties due to the strong interface between the reinforcement phase and the matrix, and through secondary strengthening effects such as dislocation strengthening. Figure shows that as the particle size of Al_2O_3 goes from micron to nanosize, there is significant decrease in the friction coefficient and wear rate of aluminum composites. In addition, the incorporation of only 10 volume percent of 50 nm sized Al_2O_3 particles to the aluminum alloy matrix resulted in an increase of the yield strength to 515 MPa. This is 15 times stronger than the base alloy, 6 times stronger than the base alloy with 46 volume percentage of 29 micron size Al_2O_3 and over 1.5 times stronger than AISI 304 stainless steel.

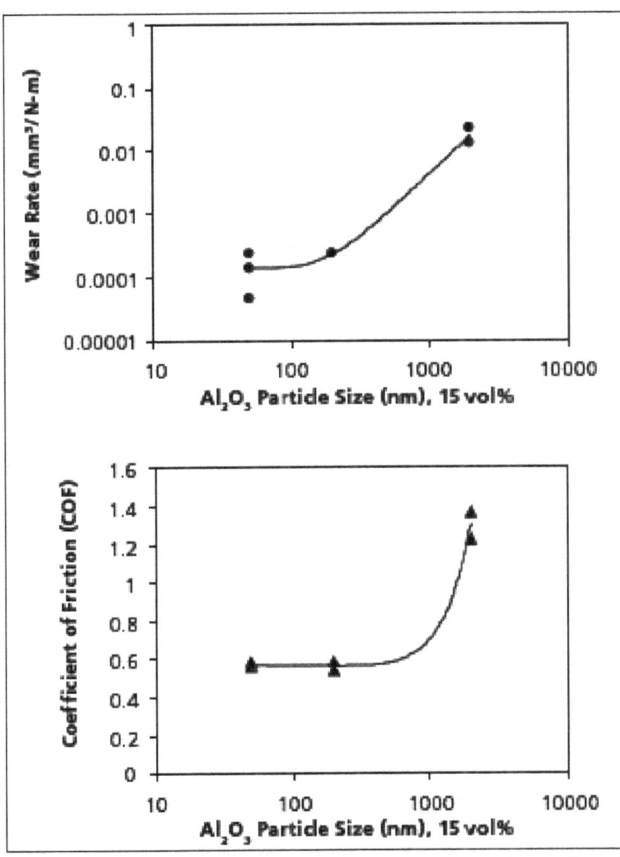

Effect of particle size on wear rate and coefficient of friction of Al-15vol% Al_2O_3 Metal-matrix composites. Both the wear rate and coefficient of friction are dramatically reduced when the particle size is reduced below 1 μm.

Reagents for non-aqueous titration		
Team	Process	Properties and Comments
Takagi et al.	Rapidly solidified nanocrystalline Al alloys and Al/SiC nanocomposites (Matrix: Al-Ni-Y-Co, Al-Si-Ni-Ce, and Al-Fe-Ti-Me, where Me: Cr, Mo, V, Zr). SiC size: #3000 to #8000	High hardness, strength and excellent wear resistance to 473 K.
ElEskandarany	High-energy ball milling of Al-SiC (2–10%) nanocomposite and consolidation using plasma activated sintering.	Fully dense bulk nanocomposites with nanocrystalline structure and uniform SiC distribution. No reaction products (Al_4C_3) or Si at the interface. Hardness and mechanical strength characterized.
Hong et al.	Deformation processing (drawing) of Cu/Nb filament nanocomposites.	At drawing strain above 10, a limiting thickness of Nb filaments of 10 nm was obtained (further deformation caused filament rupture rather than thinning). The ductility was independent of the Nb content.
Islamgaliev et al.	Cu-0.5wt% Al2O3 nanocomposites produced by high-pressure torsion technique.	High tensile strength (680 MPa) and microhardness (2300 MPa), as well as high thermal stability, creep strength and electrical conductivity were obtained.
Fekel & Mordike	Mg strengthened by 30 nm-SiC particles. Nano-SiC formed by CO2 laser-induced reaction of Si and acetylene. Micronsize Mg powder formed by gas atomization of Mg melt with Ar gas. Composite formed by hot milling followed by hot extrusion.	Tensile Strength doubled as compared to unreinforced Mg. At room temperature increase in the UTS was around 1.5 times. Significant improvement in the hardness. Milled composites exhibited lowest creep rate.
Dong et al.	CNT/Cu composite. CNT formed by thermal decomposition of acetylene. CNT mixed with Cu powder using ball mill. Ball-powder weight ratio 6:1. Pressed at 350 MPa for 5 min. Sintered for 2 h at 850 °C.	Friction coefficient reduced with increasing CNT fraction in Cu. Less wear loss with increasing CNT content. Less wear loss in Cu/CNT than Cu/C as load increased.

Recently, an aluminum alloy specimen reinforced with CNTs was synthesized using pressureless infiltration. CNTs were mixed together with aluminum and magnesium powders using ball milling and then pressed into preforms that were subsequently infiltrated by melts of the aluminum alloy matrix at 800 °C in a nitrogen atmosphere. The CNTs were observed to be well-dispersed and embedded in the matrix. Further experiments showed that up to 20 volume percent of nanotubes could be incorporated in the matrix of aluminum alloys using this process. The wear data also suggests that the presence of CNTs in the matrix can reduce the direct contact between the aluminum matrix and the steel pin and thereby decrease the friction coefficient due to the presence of carbon nanotubes. By reducing the friction coefficient, the energy loss experienced by components in frictional contact will be reduced, improving efficiency of mechanical systems. In addition, the authors believe that the incorporation of CNTs having relatively short tube lengths may allow them to slide and roll between the mating surfaces and result in a decrease of the friction coefficient. Ploughing wear appeared to be the dominant wear mechanism under the dry sliding condition. The depth of the wear grooves caused by ploughing wear decreased with increasing CNT content, suggesting that the strengthening effects of nanotubes increased the wear resistance of aluminum alloy-CNT nanocomposite.

Though nanocomposite materials exhibit ultra high-strength, there is often a trade-off that results in decreased ductility. This may be attributed to currently used processing methods that result in the formation of voids and defects, as well as to the inability of nanostructured grains or additives to sustain a high rate of strain hardening. These shortcomings, among other microstructural effects, lead to instabilities upon plastic deformation. One of the methods that has been used to overcome the lack of ductility in nanostructured materials is to incorporate nanosize dispersoids in a bimodal or trimodal microstructure. In the case of nanostructured grains, the presence of hard precipitates or nanoparticles in a Metal-matrix may act to initiate, drag and pin dislocations, reducing dynamic recovery, and thus resulting in a high strain-hardening rate that, in turn, produces larger uniform strains and higher strengths in the MMNC, along with higher ductility.

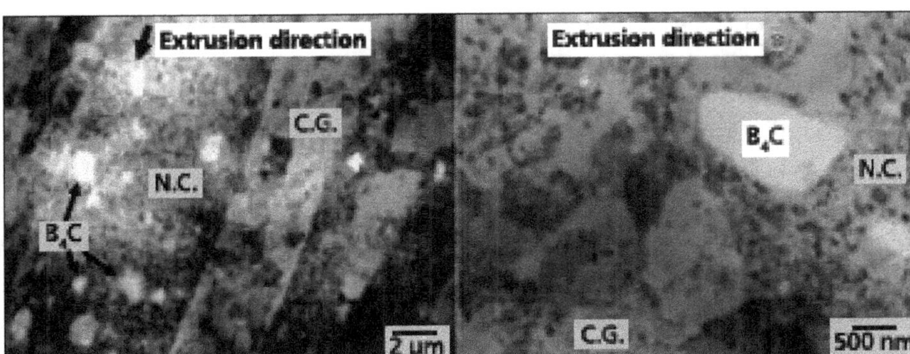

TEM micrographs showing alternate stacking of course grain size Al and nanostructured Al regions. The B_4C particles are uniformly distributed within the nanostructured aluminum exclusively.

The potential for achieving much higher strengths in combination with acceptable ductility in MMNCs has been demonstrated by E. Lavernia and coworkers. A mixture of a 5083 aluminum alloy powder and micron-sized boron carbide particles that had been coated with a powdered, nanostructured aluminum alloy were milled at cryogenic temperatures (cryomilled). This cryomilled powder was then mixed with regular 5083 alloy and consolidated and extruded to form nanocomposites with almost double the strength of the monolithic 5083 alloy. Cigar-shaped nanocomposite volumes of material that contained micron-sized B4C reinforcements were observed to be dispersed in nanograin sized aluminum matrix regions that were embedded in micro-grain sized

5083 aluminum matrix having no B_4C reinforcements. This combined bimodal and trimodal microstructure is presumed to be the source of the exceptional combination of strength and ductility in aluminum matrix nanocomposites that had micron size B_4C particles used as reinforcements. Apparently, the micron grain-sized aluminum matrix that did not contain any reinforcement, enabled dislocation motion and improved ductility. This type of combined bimodal and trimodal microstructure in MMNCs will need to be synthesized using less expensive solidification processing routes that can enable processing of large components having complex shapes. By exploring lower cost and more versatile methods to manufacture Metal-matrix nanocomposites with improved ductility, these materials are expected to become commercially viable for a variety of applications, particularly where weight savings is essential.

Applications

Metal-matrix composites with micron-size reinforcements have been used with outstanding success in the automotive and aerospace industries, as well as in small engines and electronic packaging applications. In the case of Metal-matrix nanocomposites, incorporation of as little as one volume percentage of nanosize ceramics has lead to a much greater increase in the strength of aluminum and magnesium base composites than was achieved at much higher loading levels of micron-sized additions. Such potential improvements have great implications for the automotive and aerospace, and, in particular, the defense industries due to the drastic weight savings and exceptional properties that can be achieved. Potential aerospace applications may include ventral fins for aircrafts, as well as fan exit guide vanes for commercial airline jet engines. Both components require high stiffness and strength, low weight as well as resistance to erosion from rain, airborne particulates and hail. Components used in the automotive industry where bulk nanocomposites would likely be valuable include brake system components, which require high wear resistance, and thermal conductivity, intake and exhaust valves, which require high creep resistance and resistance to sliding wear, as well as piston liners, which require high wear resistance, good thermal conductivity and low coefficient of thermal expansion. In addition, exceptionally high thermal conductivities, possible in selected nanocomposites, will find applications in thermal management applications in computers. Metal-matrix nanocomposites can be designed to exhibit high thermal conductivity, low density, and matching coefficient of thermal expansion with ceramic substrates and semiconductors, making them ideal candidates for such applications.

Carbon Nanotube Metal Matrix Composites

Carbon nanotube Metal-matrix composites (CNT-MMC) are an emerging class of new materials that are being developed to take advantage of the high tensile strength and electrical conductivity of carbon nanotube materials. Critical to the realization of CNT-MMC possessing optimal properties in these areas are the development of synthetic techniques that are (a) economically producible, (b) provide for a homogeneous dispersion of nanotubes in the metallic matrix, and (c) lead to strong interfacial adhesion between the metallic matrix and the carbon nanotubes. Since the development of CNT-MMC is still in the research phase, the current focus is primarily on improving these latter two areas.

Carbon Nanotubes Reinforced Metal Matrix Composites Production Methods

According to the new production systems, Carbon nanotubes reinforced Metal-matrix composites (CNT-MMC) may be produced in several different methods. These production methods are.

Powder metallurgy Route Technics:

- Conventional Sintering.
- Hot Pressing.
- Spark Plasma Sintering.
- Deformation Processing.
- Hot Extrusion.
- Semi-solid Powder Processing.

Electrochemical Routes(for non-structural applications):

- Electro-deposition.
- Electroless Deposition.

Thermal spraying:

- Plasma Spraying.
- HVOF Spraying.
- Cold Kinetic Spraying.

Melt Processing:

- Casting.
- Melt Infiltration.

Novel Techniques:

- Molecular Level Mixing.
- Sputtering.

- Sandwich Processing.
- Torsion/Friction Processing.
- CVD and PVD (Physical vapor deposition).
- Nanoscale Dispersion.
- Pulsed laser deposition.

Indigenous techniques:

- Such as molecular level mixing (in which CNTs are dispersed into a metal-salt bath, forming a metal-CNT precursor).

Powder Metallurgy Techniques

Sintering is one of the oldest method in production technics and used to produce density-controlled materials and components from metal or ceramic powders by applying thermal energy. Synthesis and sintering of nanocrystalline ceramic powders have attracted much attention due to their promising properties. The high active surface area of nanopowders results in lowering sintering temperature relative to coarser powders. Although low temperature sintering suppresses the grain growth, high density of interfaces and grain boundaries in nanocrystalline powders leads to accelerated grain growth during sintering.

- Conventional Sintering is the simplest method for producing CNT Metal-matrix composite compacts. The CNTs and metal powders are mixed by a process of mechanical alloying/blending and then are compressed to form a green compact, which is then sintered to get the final product. Metallic compacts are subject to oxidation as compared to ceramics and hence the sintering has to be done in an inert atmosphere or under vacuum. One major drawback of this processing route is the inability to tailor the CNT distribution within the metallic matrix.

- Microwave sintering is one of them and fundamentally different from conventional sintering. In microwave sintering process, the material is heated internally and volumetrically unlike in a conventional process where heat originates from an external heating source. Sintering cycle time for microwave sintering is much shorter as compared with the conventional sintering cycle.

- Spark plasma sintering is a quite new technique which takes only a few minutes to complete a sintering process compared to conventional sintering which may take hours or even days for the same. High sintering rate is possible in SPS since high heating rates can be easily attained due to internal heating of the sample as oppsed to external heating seen in case of conventional sintering. For conventional sintering usually a green compact needs to be prepared externally using a suitable die and hydraulic machine for applying the necessary pressure.

In SPS the powder is directly fed into the graphite dies and the die is enclosed with suitable punches. All types of materials, even those difficult to densify can be easily sintered in SPS. Due to advantage of high heating rate and less holding time, SPS can restrict the unwanted sintering reactions in highly reactive systems as opposed to conventional sintering and hence formation of undesirable product phases can be avoided.

- Semi-solid Powder Processing (SPP) is a unique method that fabricates composites materials with powder mixtures in the semi-solid states. Starting with metal-CNT powder mixture, the metal powder is heated to the semi-solid state, and pressure is applied to form the Metal-matrix composites. This method features many advantages such as simple and fast process and flexible property tailoring.

Carbon Nanotube Dispersing and CNT Breakage during Mixing

One common method to disperse the CNT into the Metal-matrix is mechanical alloying. However, many researchers reported the length reduction and damage of CNTs during mechanical alloying process.

Mechanical Properties

Carbon nanotubes are the strongest and stiffest materials yet discovered in terms of tensile strength and elastic modulus respectively. This strength results from the covalent sp^2 bonds formed between the individual carbon atoms. Multi-walled carbon nanotube was tested to have a tensile strength of 63 gigapascals (GPa). Further studies, conducted in 2008, revealed that individual CNT shells have strengths of up to ~100 GPa, which is in good agreement with quantum/atomistic models. Since carbon nanotubes have a low density for a solid of 1.3 to 1.4 g/cm^3, its specific strength of up to 48,000 $kN \cdot m \cdot kg^{-1}$ is the best of known materials, compared to high-carbon steel's 154 $kN \cdot m \cdot kg^{-1}$. CNTs are not nearly as strong under compression. Because of their hollow structure and high aspect ratio, they tend to undergo buckling when placed under compressive, torsional, or bending stress.

Comparison of mechanical properties			
Material	Young's modulus (TPa)	Tensile strength (GPa)	Elongation at break (%)
SWNT^E	~1 (from 1 to 5)	13–53	16
Armchair SWNT^T	0.94	126.2	23.1
Zigzag SWNT^T	0.94	94.5	15.6–17.5
Chiral SWNT	0.92		
MWNT^E	0.2–0.8–0.95	11–63–150	
Stainless steel^E	0.186–0.214	0.38–1.55	15–50
Kevlar–29&149^E	0.06–0.18	3.6–3.8	~2

Potential Applications

Nanonetwork

Nanonetworks are expected to expand the capabilities of single nanomachines both in terms of complexity and range of operation by allowing them to coordinate, share and fuse information. CNT Metal-matrix composites enable new applications of nanotechnology in the military technology and industrial and consumer goods applications.

Nanorobotics

Nanomachines are largely in the research-and-development phase, but some primitive molecular machines have been tested. An example is a sensor having a switch approximately 1.5 nanometers across, capable of counting specific molecules in a chemical sample. The first useful applications of nanomachines might be in medical technology, which could be used to identify and destroy cancer cells. Another potential application is the detection of toxic chemicals, and the measurement of their concentrations, in the environment. In addition CNT-MM composite will be the main material for the military robots, especially to strength the robot soldier's armours.

Future Soldier

The exhibit 'future soldier', designed for US Army.

Today's militaries often use high-quality helmets made of ballistic materials such as Kevlar and Aramid, which offer improved protection. Some helmets also have good non-ballistic protective qualities, though many do not. Non-ballistic injuries may be caused by many things, such as concussive shock waves from explosions, physical attacks, motor vehicle accidents, or falls. Another application for the future soldier is powered exoskeleton system. Powered exoskeleton, also known as powered armor, or exoframe, is a powered mobile machine consisting primarily of an exoskeleton-like framework worn by a person and a power supply that supplies at least part of the activation-energy for limb movement. Powered exoskeletons are designed to assist and protect the soldiers and officers. Currently MIT is working on combat jackets that use CNT fibers to stop bullets and to monitor the condition of the wearer.

Advanced Modular Armor Protection

Advanced Modular Armor Protection (AMAP) is modular composite armour concept, developed by the German company IBD Deisenroth Engineering. According to IBD AMAP is a 4th generation composite armour, making use of nano-ceramics and modern steel alloy technologies. AMAP is making use of new advanced steel alloys, Aluminium-Titanium alloys, nanometric steels, ceramics and nano-ceramics. The new high-hardened steel needs 30% less thickness to offer the same protection level as ARMOX500Z High Hard Armour steel. While Titanium requires only 58% as much weight as rolled homogeneous armour (RHA) for reaching the same level of protection, Mat 7720 new, a newly developed Aluminium-Titanium alloy, needs only 38% of the weight. That means that this alloy is more than twice as protective as RHA of the same weight.

Type 10 MBT consist of Nano-crystal steel (or Triple Hardness Steel), Modular ceramic composite armor, Light weight upper armor.

AMAP is also making use of new nano-ceramics, which are harder and lighter than current ceramics, while having multi-hit capability. Normal ceramic tiles and a liner backing have a mass-efficiency (E_M) value of 3 compared to normal steel armour, while it fulfills STANAG 4569. The new nano-crystalline ceramic materials should increase the hardness compared to current ceramics by 70% and the weight reduction is 30%, therefore the E_M value is larger than 4. Furthermore, the higher fracture toughness increases the general multi-hit capability. Some AMAP-modules might consist of this new ceramic tiles glued on a backing liner and overlaid by a cover, a concept which is also used by MEXAS. Lightweight SLAT armour is also part of the AMAP family.

Leopard 2SG of the Singapore Army upgraded with AMAP Composite Armour by IBD & ST Kinetics.

Nano Armour

The TK-X (MBT-X) project, the new Type 10 main battle tank design use composition of modular components of nano-crystal steel (nor Triple Hardness Steel), Modular ceramic composite armor, partly reinforced MMC and Light weight upper armor.

Materiomics

Materiomics is defined as the study of the material properties of natural and synthetic materials by examining fundamental links between processes, structures and properties at multiple scales, from nano to macro, by using systematic experimental, theoretical or computational methods and refers to the study of the processes, structures and properties of materials from a fundamental, systematic perspective by incorporating all relevant scales, from nano to macro, in the synthesis and function of materials and structures. The integrated view of these interactions at all scales is referred to as a material's materiome.

Materiomics includes the study of a broad range of materials, which includes metals, ceramics and polymers as well as biological materials and tissues and their interaction with synthetic materials. Materiomics finds applications in elucidating the biological role of materials in biology, for instance in the progression and diagnosis or the treatment of diseases. Others have proposed to apply materiomics concepts to help identify new material platforms for tissue engineering applications, for instance for the de novo development of biomaterials. Materiomics might also hold promises for nanoscience and nanotechnology, where the understanding of material concepts at multiple scales could enable the bottom-up development of new structures and materials or devices, including biomimetic and bioinspired structures.

Nanotough

Nanotough is to obtain a deeper understanding of the interfacial structure of nanocomposites within a polyolefin matrix and thus use nanoparticles like nanoclay to turn upside down the construction of a number of well-known products, where today metals or plastics are used in for example cars or aircraft. The project will enable realization of the great performance potential of these materials through development of novel multiphase and hybrid nanocomposites.

Nanotough project aims to improve the stiffness of polyolefin nanocomposites while not only maintaining but also improving the toughness of the matrix considerably. The technical objective is to optimize and, through novel interface design, to develop new cost effective hybrid (nanofiller-fibre) nanocomposites as an alternative to heavily filled polymers and expensive engineering polymers and fulfill industry requirements for high performance materials in high tech applications.

Electrodeposition of Metal Matrix Nanocomposites

Research into the preparation of nanocomposite coatings, by electrochemically codeposition of fine particles with metal from electrolytic solutions, has been investigated by numerous authors.

Materials are considered nanosized when one of the components dimensions are in the nanometer scale, with typical dimensions smaller than 100 nm. A variety of nanosized particles, ranging from 4 nm to 800 nm diameters, have been successfully incorporated into metallic electrodeposits. These include oxides such as Al_2O_3, ZrO_2, TiO_2 and Cr_2O_3 or carbides like SiC, WC, TiC or nitrides like Si_3N_4.The metals mostly used are copper and nickel however other metals like zinc, silver and, alloys have been used. The most studied system has been the Ni-SiC due to its potential technological applications. Therefore, considering the variety of metals, which can be electrodeposited, electrochemical composite deposition enables the production of a wide range of composite materials that compared to the pure metal coatings have improved physical and (electro)chemical properties. For example, Cualumina and Ni-alumina composite coatings exhibit ultimate tensile strengths, and lower ductility than pure copper or nickel deposits.

The first application of electrochemically deposited composites dates back to the beginning of the twenty century. Sand particles held by a nickel matrix were utilized as anti-slip coatings on ship stairs. During the 1970s and 1980s, investigations were focused on the need to produce coatings with enhanced mechanical, corrosion and tribological properties. During the 1990s, new areas such as electrocatalysts and photoelectrocatalysts production were emerging, associated with an increasing interest on the particles size. This is due to the higher activity of the semiconducting nanoparticles in comparison with the macro sized particles related to changes on the optical bad gap and surface area. Concerning photoelectrochemical activity, the principal application of these materials is strongly linked to the degradation of organic molecules and depends on the high surface-to-volume distribution of the nanoparticles. Although the use of metallic matrix composite electrodes, on the organic pollutant degradation, is still rare, some studies indicate that the metallic matrix has a positive influence on the semiconductor photocatalytic activity. The authors of this chapter are developing, with success, the preparation of photoelectroactive thin films consisting of nanoparticles (TiO_2) dispersed on a metallic matrix (Zn or Zn alloys) suitable to be used as electrode materials for degradation of pharmaceuticals from simulated sewage water.

The amount of incorporated particles is the key parameter for the success of Metal-matrix composite applications, since it largely determines the composite properties such as wear resistance, high-temperature corrosion protection, oxidation resistance and self-lubrication when compared with the corresponding values for pure metal or alloys deposits. In addition, uniform distribution of codeposited particles within the Metal-matrix was found to be another crucial parameter. Nevertheless it must be highlighted that the Metal-matrix morphological and structural characteristics are strongly affected by the particles presence. Furthermore the co-deposition of a sufficient amount of nonagglomerated particles should lead to production of harder and more resistant coatings. Particle-reinforced composite coatings based on nickel and alumina are being applied in different technological fields with high demands on friction and corrosion resistance. Lekka et al. show that the co-deposition of SiC nanoparticles leads to a more noticeable grain refinement and, as a consequence, the nanocomposite deposits present a very high microhardness, 61% higher than pure copper deposits, and an increase of 58% of the abrasion resistance.

Future applications of these materials depends on the ability to produce them with controlled composition and properties, using inexpensive and reliable techniques.

Electrodeposition method meets some of these requirements, since it is an economical and versatile technique compared to other preparation techniques. Moreover, it is quickly scaled up to industrial production, offering an inexpensive method to produce large area samples. These are some of the reasons that make this method so popular. In addition, the low processing temperature (around room temperature) minimizes interdiffusion or chemical reaction between the substrate and coating species. The film thickness can be accurately controlled by monitoring the consumed charge and the composition can be tailored by the electrical applied profile and bath composition. The occlusion electrodeposition, or electrochemically co-deposition method, where the particles incorporation occurs simultaneously with the metal ions reduction, uses a precursor bath loaded with the particles to be occluded. This technique is widely used to obtain Metal-matrix composites due to its ease of preparation, low cost and versatility. The concentration of particles suspended in solution varies from 2 up to 200 g/L producing composites with typically 1-10 vol.% of embedded particles.

With the increasing availability of nanoparticles, the interest of the low-cost and low-temperature composite electroplating is continuously growing, with the major challenge being the achievement of high co-deposition rates and homogenous distribution of the particles in the metallic matrix.

Fundamentally, the electrocodeposition process and hence the structure, the morphology and the properties of the composite coatings is affected by the electrodeposition parameters like the electrolysis conditions (composition and agitation of the electrolytic bath, presence of additives, temperature, pH), the electrical profile and the particle properties (type, size, shape, surface charge, concentration and dispersion in the bath).

Theoretical Models of Metal Matrix Composite Electrodeposition

In a simple way, the particle incorporation in a Metal-matrix could be described as a fourstep process: (1) Formation of surface charge on particles in suspension, (2) particle mass transfer from the bulk of the suspension to the electrode surface, (3) particle-electrode interaction (4) particle incorporation and irreversible entrapment simultaneously with the growing metal layer.

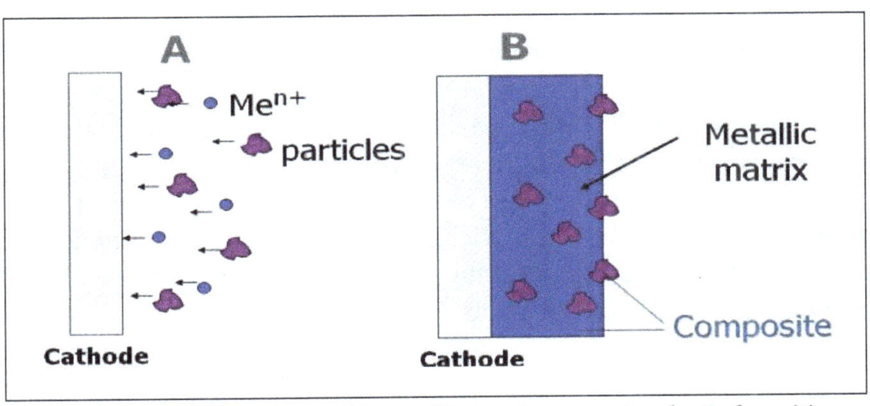

Some of the steps involved in the metallic matrix composite electrodeposition.

Most of the electrochemical co-deposition mechanisms for the dispersion of inert particles into metallic coatings have been developed for micron-sized particles. The first model developed was the Guglielmi's model which is based on two steps process involving a loose adsorption and strong adsorption of the particles. The first step is a loose physical adsorption of the particles on the cathode, with a high degree of coverage, and without discharge of the electro-active ions adsorbed on the particles. The fractional coverage follows a Langmuir adsorption isotherm. The second step is the strong electrochemical adsorption of the particles, caused by the applied electrochemical field, accompanied by the discharge of the electro-active ions. Both steps take place at the same time all over the cathode surface. If a particle is strongly adsorbed on the cathode, it will be embedded in the growing metal layer by the electrodeposition of free solvated electro-active ions from the plating bath.

From this model the volume fraction of incorporated particles, α, can be mathematically expressed by:

$$\frac{\alpha}{1-\alpha} = \frac{zF\rho_m V_o}{M_m i_o} . e^{(B-A)\eta} . \frac{kc_{p,b}}{1+kc_{p,b}}$$

where M_m and ρ_m are the atomic weight and the density of electrodeposited metal respectively, i_o the exchanging current density, z the valence of the electrodeposited metal, F the Faraday constant, η the electrode reaction overpotential, $C_{p,b}$ the particle concentration in the bulk electrolyte and k the Langmuir isotherm constant, mainly determined by the intensity of interaction between particles and cathode. The parameters V_o and B are related to particle deposition, and both play a symmetrical role with the parameters i_o and A related to metal deposition. The validity of Guglielmi's model has been verified with different deposition systems, such as SiC and TiO_2 particles with nickel from sulfamate bath, or α-Al_2O_3 particles with copper from a copper sulfate plating bath.

However the model neglects hydrodynamics, particle size and ageing effects what is its main drawback.

The model developed by Celis uses probability concept to describe the amount of particles that can be incorporated at a given current density and assumes that five steps are involved into the incorporation of the particles: (1) formation of an ionic cloud around the particles; (2) transport of the particles by convection to the hydrodynamic boundary layer; (3) transport of the particles by diffusion to the cathode; (4) free ions and electro-active ions adsorbed on the particles are adsorbed at the cathode, and (5) electrochemical reduction of the absorbed ions at the cathode with the embedding of the particles into the growing metallic matrix.

This model was satisfactory used for the Cu-Al_2O_3 and Au-Al_2O_3 systems. Later on Fransaer proposed another model that was developed to non-Brownian particles, particles with a diameter bellow 1 μm, involving two steps: (1) reduction of metal ions (described by Butler-Volmer expression) and (2) co-deposition of particle (described by trajectory expression). This model has been used successfully on the qualitatively explanation of experimental data for Zn-polystyrene composite depositions. From a mechanistic point of view, this trajectory model offers a very good description for getting a better insight into composite deposition. Nevertheless the particle-electrode interaction forces and their relative importance remain a point of discussion. Another contribution in this area is the model of Vereecken that takes into account the particles kinetics and residence time at the electrode surface. The transport of particles to the surface is controlled by convective-diffusion. The influence of particle gravitational force and hydrodynamics is accounted for various current densities. It is valid only when the particle size is smaller than the diffusion layer thickness. The authors show that incorporation of 300 nm Al_2O_3 particles into nickel films can be well described by this model. More recently Lee and Talbot, proposed a model to predict the amount of nanoparticle incorporation in a kinetically as well as in the mass transfer limited region of the electrochemical deposition which agrees well with the experimental data of the Cu-Al_2O_3 system.

Currently, the models used to describe the inclusion of particles are restricted to specific conditions, so empirical laboratory trials remain very important. Future models, describing the co-deposition process, will require special attention to interactive variables such as the nanoparticle characteristics (composition and crystallographic phase, size, density, and shape) and the operating electrolysis parameters. The validity of the various theoretical models, underlying particle incorporation in a Metal-matrix still requires attention, since the electrochemical co-deposition is still not fully understood.

Particle Characteristics

Nanocomposite coatings have been studied increasingly in the last 10 years. However the results are often contradictory and inconsistent and they are, therefore, difficult to compare with micro-composites. The mass transport of the inert particles towards the cathode surface is a crucial factor in determining the extent of their occlusion within the growing metallic matrix. The particle electrode interaction depends on the particle surface properties, which are determined by the particle type, bath composition, pH and metal surface composition. For some systems like Ni-SiC, the two-step co-deposition mechanism proposed by Guglielmi is valid. In addition the results show that the ultra-fine SiC is more difficult to co-deposit than the coarse SiC, and the rate determined step is controlled by the transferring process of loose adsorption to strong adsorption. With the same quantity of powder in the bath, the embedded micro-powder content is about 25–30 wt % while the nano-powder content is always less than 1 wt %.

The smaller the particle the most important are colloidal forces (van der Waals, electrostatic attraction/repulsion and hydrophobic/hydrophilic interaction forces). The most difficult to quantify are the hydrophobic/hydrophilic interaction forces, as no general theory describing such forces may be found in the literature.

Particles dispersed in a continuous electrolyte solution are in constant Brownian motion. When two particles approach one another, the energy existing between the particles determines whether the particles will separate or agglomerate. Generally, particle agglomeration occurs when larger attraction than repulsion energy exists between them. The magnitude of the net forces involved, in producing an agglomerated structure, clearly depends on the conditions and the nature of the system. The knowledge of the interfacial region structure is an important factor in order to understand the dispersion stability of solid particles within the electrolyte.

At higher ionic strength of the electrolyte or lower particle charge, the particles may agglomerate irreversibly. If the ionic strength increases, the agglomeration of the particles increases too. However if the state of agglomeration is reversible it can be easily turned into the dispersed state by mechanical stirring. Repulsive forces can be tuned via experimental conditions as pH, ionic strength, etc. The zeta potential, which is a quantitative measure for the particle surface charge, gives an indication of the stability of the colloidal system. As consequence, a higher zeta potential induces a lower degree of particle agglomeration producing a higher concentration of non-agglomerated particles in the electrolyte. In electrocodeposition the stability of the colloid dispersion is mandatory, since the particles shall be preferentially incorporated in the form of individual particles in order to improve the particle incorporation and as consequence the composite properties. Simunkova et al. show, for the nickel composites, that as electrostatic repulsion within the dispersion increases, particle agglomeration slows down and the concentration of non-agglomerated particles rises resulting in an enhanced amount of particles incorporated in the electroplated Ni layers. Effective particle dispersion in the electrolyte solution creates more opportunities for loose particle adsorption onto the electrode. Smaller agglomerated particle groups have a higher attraction force from the electric field that produced effective adsorption.

The zeta potential of nano- and micro-scaled particles is influenced by many factors, such as the source of particles (variable preparation and stabilization), the treatment with different surfactants, the electrolyte concentration (ionic strength), the particle morphology and size, the pH of the solution, and the state of hydration.

The particles always interact with the electrolyte and therefore chemical and physical adsorption of electrolyte ions onto the particle occurs. This adsorption and the initial particle surface composition determine the particle surface charge, which induces a double layer of electrolyte ions around the particle. In electrolytes double layers play a major role in the interactions between particles and also between particles and the electrode. In aqueous media, the oxide particles, due to protonation– deprotonation of the superficial groups can change their surface charge depending on the solution pH according to:

$$M-OH+H^{+} \rightarrow -M-OH2^{+}$$

$$-M-OH+OH^{-} \rightarrow -M-O^{-}+H_{2}O$$

For example the TiO_2 point of zero charge (pzc) is ca 6.0, thus for pH values lower than 6.0, the positively charged TiO_2 particles would be expected to be electrostatically attracted to the cathode surface, thus assisting in their solution transport via migration. On the contrary, at solution pH values higher than the pzc, the TiO_2 particles would be negatively charged and thus repelled from the cathode surface. For the Ni-TiO_2 system, the experimental data revealed that co-deposition of TiO_2 nanoparticles is favoured at low pH values and current densities, implying that there is a plentiful adsorption of H+ on the titania surface. As the particle surfaces become positively charged they will be strongly adsorbed on the cathode leading to an enhanced electrolytic co-deposition.

Nevertheless, in some published works, it is pointed out the electro deposition of negatively charged particles into Ni matrix with success. These findings are in contradiction with the traditional theory of particles transport towards the cathode by means of electrophoresis. This might be correlated to the absolute value of the zeta potential. This factor is very important for the degree of incorporation and not only the polarity of the zeta potential.

According to the model developed by Bund and Thiemig, negatively charged particles are preferentially attracted by the positive excess charge in the electrolytic part of the electrode electrical double layer. When the particle comes closer to the electrode, the shell of adsorbed ions on the particle is stripped off, within the electric double layer and the particle becomes incorporated into the growing metal layer.

As general accepted the transport of the particles to the cathode surface occurs by electrophoresis, mechanical entrapment, adsorption and convective-diffusion mechanisms. Electrolyte agitation is usually necessary to maintain the particles in suspension and to transport the particles to the electrode surface. Vaezi et al. have verified that increasing the stirring rate up to 120 rpm causes the increase of the percentage of incorporated SiC nanoparticles but when the stirring rate is too high, a decreasing trend of the weight percent is observed, principally caused by the collision factor. At a high-stirring rate, because of the turbulent flow in the bath, the SiC nanoparticles on the cathode surface are washed away and thus the SiC nanoparticles percentage in the composite coating decreases. It was assumed that increasing the stirring rate, increases the forces acting on the nanoparticles, resting on the cathode surface, decreasing the weight percentage of the SiC nanoparticles in the composite coating.

The co-deposition of nanoparticles during electrochemical deposition of a metallic film depends on the rate of the metal deposition and on the flux of particles to the film surface which is strictly related with the applied electric profile. The growth rate of the metal film is determined by the deposition current density.

Generally, the strong adsorption is the rate-determining step of the electrodeposition process because it is more difficult to occur than the loose adsorption. The strong adsorption can be promoted by high overpotential corresponding to increasing current density. As a result, the particle content in the coatings increases with increasing current density. Nevertheless, at higher current density, the improvement of metal deposition rate caused by the increase in current density exceeds the promotion effect of particle incorporation, which leads to the decrease in the particle content.

The relationship between cathodic current density and incorporation rate, for different particle concentrations in the bath, has been studied by Gay et al.. It was concluded that for low current densities (0.5 A dm^{-2}), the incorporation rate is very weak <2% and for higher current densities (1.5 to 2 A dm^{-2}), the coatings appear rough. In these latter cases, the incorporation rate is still very weak (approx. 1%) whatever the plating bath concentration is.

Studies on the Ni-TiO_2 system done by Aal et al. show that the wt. % of co-deposited TiO_2 increases with current density. According to the Stokes model, assuming that the colloidal particle is sphere-shaped, the electrophoretic velocity (V_E) under an electric field (E) can be expressed by:

$$V_E = \mu_E E = \frac{q}{6\pi\eta r} E$$

where μ_E is the electrophoretic mobility, q the particle charge, r the particle radius, and η the viscosity of the suspension. Considering this, the increase of the current density accelerates the electrophoretic velocity of the TiO_2 particles and increases the Coulombic force between the Ni^{2+} adsorbed on the particles and the cathode which consequently increases the TiO_2 content in the Ni deposit. Similar behavior was revealed for the Ni-SiC composites.

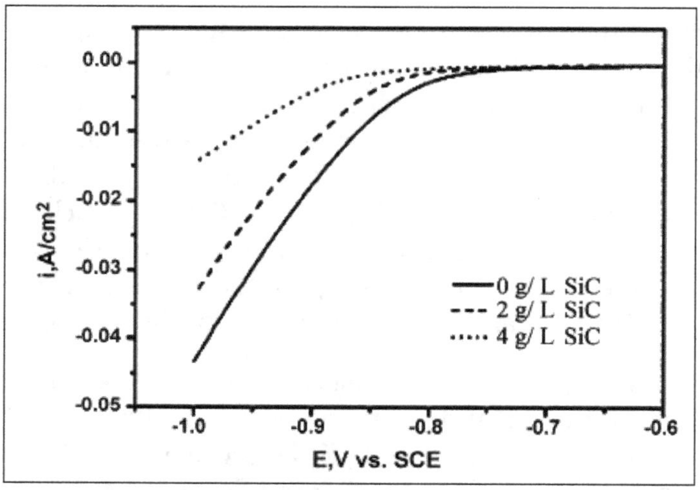

Cathodic polarization curves for the deposition of Ni–SiC at different concentrations of SiC nanoparticles in the bath.

Otherwise, it has been observed that the amount of embedded SiC particles increases with both increasing concentration of suspended SiC particles and additives' presence in the electrolyte. Indeed the particle loading in suspension is a very important parameter. At low loadings, co-deposition is limited by the supply of particles to the electrode. As the particle loading increases, so does the incorporation level. But at the highest loadings (beyond which particle settling becomes significant), the increase in incorporation is not proportional to the increase in loading and a constant value is attained. In many cases the enhanced performance of the composite film is mainly caused by a change in the Metal-matrix growth mode or crystallite size and not so much by the presence of the particles themselves. In general the crystallite size of the Metal-matrix decreases due to the presence of the nanoparticles in the electrolyte. For instance the crystallite size of the nickel matrix was reduced from 115 to 30 nm due to the presence of 120 g/L alumina nanoparticles. Some studies have focused on the particle effect on the metal reduction rate and the conclusions are in somehow contradictory. Few works have shown that there is no change in the metal deposition rate with particle addition to the electrolyte. Lozano-Morales et al. show, for Cu-Al_2O_3 nanocomposites, that, at a particle loading of 12.5 g/L, there is no change in the kinetic behaviour whatever the rotation rate is. However, the Cu reduction is inhibited, in kinetically limited region, when 39 g/L particles are added. Under mass-transfer control, the limiting current density remained the same at low particle loadings and was enhanced for higher particle concentrations.

The cathodic polarization curves for the deposition of the Ni–SiC composites containing different concentrations of SiC nanoparticles are shown in figure.

It is clear that the addition of SiC nanoparticles to the electrolyte, shifts the reduction potential of nickel towards larger negatives values, but the slope of the curve keeps unchanged. This shift is attributed to a decrease in the copper cathode active surface area, owing to the adsorption of the SiC nanoparticles, and may also be related to the decrease in the ionic transport by the SiC nanoparticles, which does not significantly affect the electrochemical reaction mechanism. By contrast Benea et al., found for the same system that the addition of SiC nanoparticles displaces the nickel reduction curve to more positive potentials. The shift in the reduction potential was attributed to an increase in the active surface area due to the adsorbed particles on the nickel cathode and to a possible increase in ionic transport by the nanoparticles.

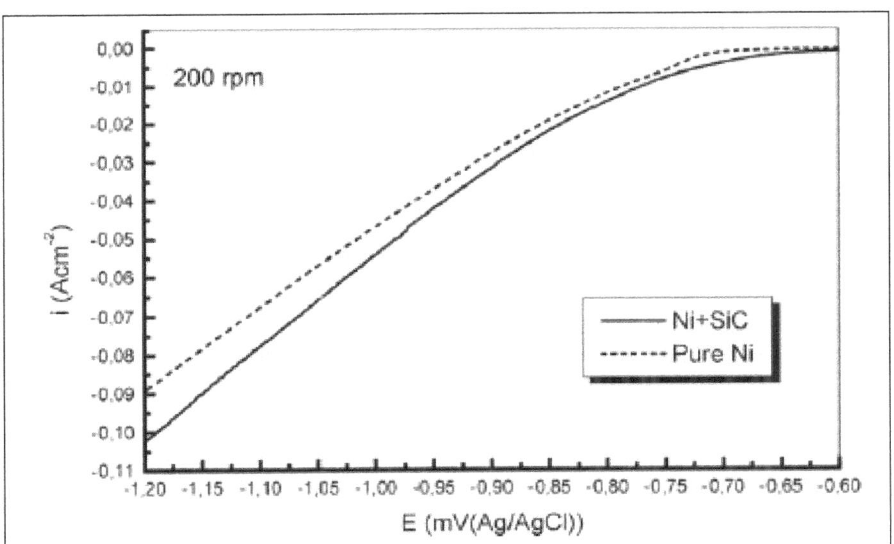

Cathodic potentiodynamic diagrams for electrodeposition of nickel in presence and absence of SiC nanoparticles. Sweep rate is 0.5 mV/s.

The lack of reproducibility shown by these two results could be correlated to the different deposition conditions used namely, cathode nature, particle size, solution pH and temperature.

In addition, the metal deposition rate, determined by average current density, have a huge influence on the particle content of the coatings.

Pulse Plating Methodology

Various electrodeposition techniques namely pulsed current, pulsed reverse current and pulsed potentiostatic can be employed to improve the incorporation of nanoparticles into metal matrix.

Several published studies have proved that the application of pulsed current (PC) technique in nickel electroplating results in the production of composite deposits with enhanced mechanical properties, higher percentages of incorporation and more uniform distribution of the particles in the metallic matrix than those attained by direct current (DC) technique.

Electrochemical deposition by pulsed current, is a very versatile method, since apart from the type, shape and amount of dispersed particles in the electrolyte, the composite properties can be optimized by the variation of the electrodeposition parameters such as current density, duty cycle and pulse frequency. The formation of deposits with desired composition, structure and porosity can be promoted by an adequate choice of these parameters.

The pulsed current parameters are the pulse length ton, the time between two pulses toff, the pulse height ip and the average current density iA, defined as:

$$i_A = (i_P \cdot t_{on}) / (t_{on} + t_{off})$$

The duty cycle is the ratio of pulse length (on-time) and the sum of on- and off-time and frequency is the inverse of the sum of t_{on} and t_{off}. The imposed pulse height causes the depletion of ions near the cathode. During t_{off}, ions migrate to the interfacial region and when t_{on} comes after the end of the first cycle, the repeating of the cycles occurs.

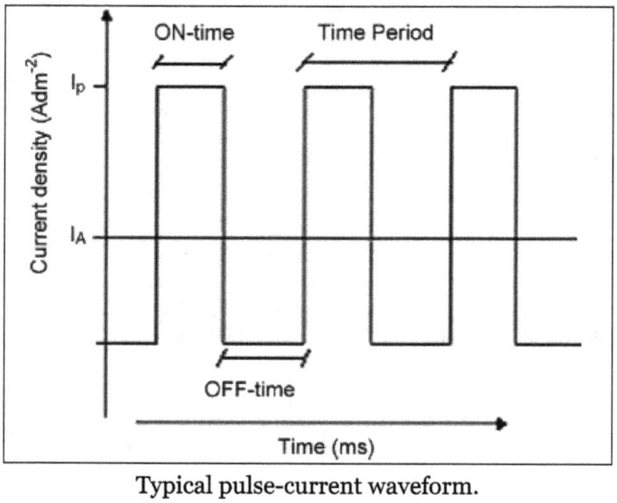

Typical pulse-current waveform.

By modifying these parameters changes on the cathodic overvoltage may occur, which affects the rate and activation energy of nucleation. At a given average current density, a decrease on the metal crystallite dimensions, with the decreasing of t_{on}, when followed by a short t_{off}, have been reported. These results have been assigned to an increase of the number of crystal nucleus formed. In addition longer toff promotes the arriving of more particles near the cathode and consequently a higher number of particles are incorporated. Therefore pulse plating techniques are very important for the tailoring of nanocomposites. In the case of Ni- SiC co-deposition, the application of pulsed current techniques results in the production of composite coatings with higher percentages of incorporation, reduced crystallite sizes and a more uniform distribution of SiC particles in the Ni matrix than those achieved under direct current regime.

In the pulsed-reverse current technique (PRC) a stripping time is introduced into the plating cycle, during which the metal surface protrusions selectively dissolves, what ensures a more uniform deposit. Comparing the PRC with the direct current method: (1) the amount of nanoparticles in a metal deposit can be enhanced, (2) a lower concentration of nanoparticles in the electrolyte solution can be used and (3) selective entrapment of nanoparticles with similar sizes can be achieved. These features make this technique most adequate for the preparation of composites, enabling an increase of the particle incorporation in the Metal-matrix Studies on the preparation of Zn-TiO_2 nanocomposite using the PRC technique have shown an improvement on the TiO2 content in the nanocomposite.

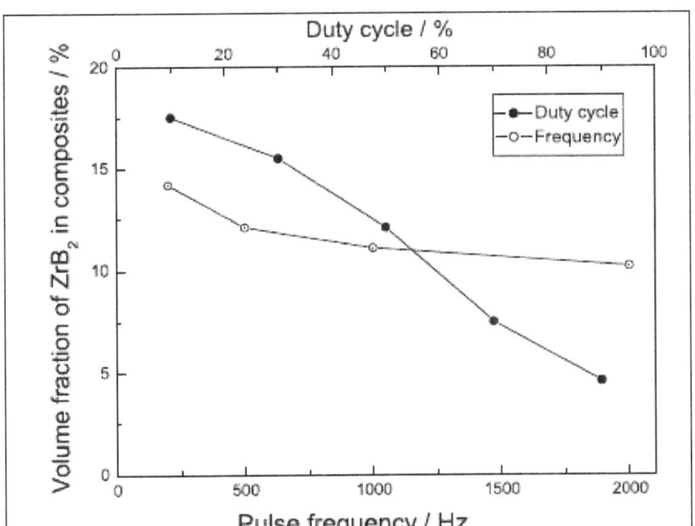

Effect of the pulse frequency with 50% duty cycle and the duty cycle with 500 Hz frequency on the volume fraction of embedded ZrB2 particles at an average current density of 5 A dm-2.

In general, for pulse plating and pulse reverse plating, high amounts of incorporated particles were found at low values of the average current density. On the other hand, various authors assumed that the duty cycle and pulse frequency affect the volume percentage of incorporated particles in the coatings. It is interesting to note that the influence of the duty cycle on the amount of incorporated particles is much bigger than the effect of the pulse frequency. These results indicate that when the pulse off-time is longer, i.e. at low duty cycles, more particles are incorporated into the coating, due to continuous agitation, with renovation of the bath composition near the cathode. It seems that the duty cycle, has a greater influence on the amount of incorporated particles than simply the length of the off-time, indicated by pulse frequency. This effect is illustrated in figure for the Ni-ZrB_2 system. Moreover, in a study of pulse plating nickelalumina composite coatings, Bahrololoom and Sani investigated both the effect of the duty cycle and frequency on the composites hardness. They concluded that decreasing duty cycle and frequency together, increased the composites' wear resistance in the same way that pulse parameters influence their hardness.

At constant average current density, a lower duty cycle means a higher peak current density. Consequently, as showed previously, the strong adsorption of particles is improved by high overpotentials. At high current densities, metallic ions are transported faster than the particles, which are transported by the mechanical agitation. Hence, the co-deposition of particles becomes particle-transfer controlled. Besides, at higher current densities the particles do not have enough time to be loosely adsorbed on the electrode surface and, as a consequence, a lower level of incorporation of the particles is obtained. In addition, at higher current densities, the hydrogen bubbles generated at the cathode surface, tend to attract the particles and prevent them from co-deposition in the metal matrix.

References

- Suk-Joong, L. Kang Sintering -Densification, Grain Growth and Microstructure, Elsevier Butterworth-Heinemann, 2005, ISBN 0-7506-6385-5, p.3
- light-weight-metal, material-matters, technical-documents: sigmaaldrich.com, Retrieved 12 May, 2020
- Patel, G. M.; Patel, G. C.; Patel, R. B.; Patel, J. K.; Patel, M. (2010). "Nanorobot: A versatile tool in nanomedicine". Journal of Drug Targeting. 14 (2): 63–67. doi:10.1080/10611860600612862. PMID 16608733
- Electrodeposition-of-metal-matrix-nanocomposites-improvement-of-the-chemical-characterization-techni, advances-in-nanocomposites-synthesis-characterization-and-industrial-applications: intechopen.com, Retrieved 24 April, 2020
- Wang, J. (2009). "Can Man-Made Nanomachines Compete with Nature Biomotors?". ACS Nano. 3 (1): 4–9. doi:10.1021/nn800829k. PMID 19206241

PERMISSIONS

All chapters in this book are published with permission under the Creative Commons Attribution Share Alike License or equivalent. Every chapter published in this book has been scrutinized by our experts. Their significance has been extensively debated. The topics covered herein carry significant information for a comprehensive understanding. They may even be implemented as practical applications or may be referred to as a beginning point for further studies.

We would like to thank the editorial team for lending their expertise to make the book truly unique. They have played a crucial role in the development of this book. Without their invaluable contributions this book wouldn't have been possible. They have made vital efforts to compile up to date information on the varied aspects of this subject to make this book a valuable addition to the collection of many professionals and students.

This book was conceptualized with the vision of imparting up-to-date and integrated information in this field. To ensure the same, a matchless editorial board was set up. Every individual on the board went through rigorous rounds of assessment to prove their worth. After which they invested a large part of their time researching and compiling the most relevant data for our readers.

The editorial board has been involved in producing this book since its inception. They have spent rigorous hours researching and exploring the diverse topics which have resulted in the successful publishing of this book. They have passed on their knowledge of decades through this book. To expedite this challenging task, the publisher supported the team at every step. A small team of assistant editors was also appointed to further simplify the editing procedure and attain best results for the readers.

Apart from the editorial board, the designing team has also invested a significant amount of their time in understanding the subject and creating the most relevant covers. They scrutinized every image to scout for the most suitable representation of the subject and create an appropriate cover for the book.

The publishing team has been an ardent support to the editorial, designing and production team. Their endless efforts to recruit the best for this project, has resulted in the accomplishment of this book. They are a veteran in the field of academics and their pool of knowledge is as vast as their experience in printing. Their expertise and guidance has proved useful at every step. Their uncompromising quality standards have made this book an exceptional effort. Their encouragement from time to time has been an inspiration for everyone.

The publisher and the editorial board hope that this book will prove to be a valuable piece of knowledge for students, practitioners and scholars across the globe.

INDEX

www.ingramcontent.com/pod-product-compliance
Lightning Source LLC
LaVergne TN
LVHW080622210926
848485LV00003B/196